Forests of Refuge

Forests of Refuge

DECOLONIZING ENVIRONMENTAL GOVERNANCE IN THE AMAZONIAN GUIANA SHIELD

Yolanda Ariadne Collins

UNIVERSITY OF CALIFORNIA PRESS

University of California Press
Oakland, California

Library of Congress Cataloging-in-Publication Data

Names: Collins, Yolanda Ariadne, author.
Title: Forests of refuge : decolonizing environmental governance in the Amazonian Guiana Shield / Yolanda Ariadne Collins.
Description: Oakland, California : University of California Press, [2024] | Includes bibliographical references and index.
Identifiers: LCCN 2023030774 (print) | LCCN 2023030775 (ebook) | ISBN 9780520396067 (cloth) | ISBN 9780520396074 (paperback) | ISBN 9780520396081 (ebook)
Subjects: LCSH: Reducing Emissions from Deforestation and Forest Degradation (Program) | Forest conservation—Guiana Highlands. | Greenhouse gas mitigation—Guiana Highlands. | Forest degradation—Control—Guiana Highlands. | Deforestation—Control—Guiana Highlands. | Environmental policy—Social aspects—Guiana Highlands.
Classification: LCC SD414.A43 C65 2024 (print) | LCC SD414.A43 (ebook) | DDC 333.75/16098—dc23/eng/20231120
LC record available at https://lccn.loc.gov/2023030774
LC ebook record available at https://lccn.loc.gov/2023030775

Manufactured in the United States of America

33 32 31 30 29 28 27 26 25 24
10 9 8 7 6 5 4 3 2 1

For a fairer, more equitable world

Contents

List of Figures and Tables

FIGURES

TABLES

Acknowledgements

I wish to acknowledge all those people and institutions who shaped my academic and personal journey thus far, without whom this book would simply not have been possible.

I am profoundly grateful, first and foremost, to my family who unwaveringly supported my academic endeavors and instilled in me the confidence that I was capable of doing all I set out to do. I am grateful for parents who scarcely questioned my excursions into poorly accessible places and distant cities, and who accompanied and supported me on these journeys as best they could. I am grateful also to my siblings and wider network of friends, family, and colleagues who listened to my many tales and travails along the way and reminded me in moments of crisis (often imagined and of my own making) that things would work out as they should.

I am indebted to the universities that financially supported my graduate studies. Scholarships from the University of Westminster, London, UK, and from Central European University, Budapest, Hungary / Vienna, Austria, opened doors to me that would have otherwise remained closed. These universities provided the space and resources that enabled my curiosity to unfold.

I am deeply appreciative of those with whom my doctoral research shared an institutional origin. Of those, I am especially thankful to Dr. Noémi Gonda for bravely shining a light where I was at times too timid to go, to Dr. Amanda Winter for living a life of conviction that continues to inspire me, to Dr. Anna Ruban whose strength and determined self-confidence motivate me, and to Ágnes Kelemen whose brilliant mind and commitment to being the change she wants to see in the world still encourage me to do the same. I am also thankful for the inspiring company and demonstrated convictions of other doctoral students in the Department of Environmental Sciences and Policy at Central European University with whom I crossed paths along the way. I am indebted to Dr. Alexandra Oanca, without whom the experience of reading critical anthropology of development monographs would certainly have been less enjoyable, and to the Department's Administrator, Györgyi Puruczky, for sharing her time, thoughts, and life experiences with me throughout my time at CEU. *Köszönöm.*

I am grateful to my PhD advisors—Drs. Guntra Aistara, Prem Kumar Rajaram, and Robert Fletcher, all of whom played vital supportive roles throughout my time grappling with the questions underpinning my doctoral research. I am especially indebted to Dr. Robert Fletcher and Professor Dan Brockington, my doctoral defense opponent, without whom I would not have chosen this scholarly path, one that I certainly do not (yet) regret; and to Professor Bram Büscher who forcefully reminded me of the likely gendered origins of my past self-doubt. Their support continues to be indispensable to me as I question, learn, and unlearn the lessons of the modern world around me.

I am eternally grateful to the individuals who worked at the many institutions that hosted me during my data collection phase as I sought access to information and interviewees in my research sites. I am thankful to my respondents, who shall remain unnamed but who took their time and energy to feed, house, and accommodate a stranger travelling to unfamiliar places, both literally and figuratively. These people owed me nothing. Yet, they were kind and generous in ways that continue to inspire me and my work. I deeply hope that this book redounds to their benefit in some way. *Bedankt.*

I am especially thankful to Gábor Kende and his family, without whose support, love, and hospitality this book simply would not have been.

I am grateful to the 2018–20 Postdoctoral Fellows of the Institute for Cultural Inquiry Berlin, who taught me new radical ways of seeing the world, while pushing against my own assumptions and biases. I am forever indebted to Drs. Christoph Holzhey, Manuele Gragnaloti, Arnd Wedemeyer, and Claudia Peppel for having believed in my ideas enough to offer me a prestigious fellowship in an open and vibrant environment that stimulated incisive thinking and scholarship while broadening my horizons. I am especially grateful to Christoph's vision and work towards a more interdisciplinary world where scholars of different backgrounds and disciplines work towards understanding each other in occasional discomfort, rather than eternally separated in disciplinary confinement. *Danke.* I am grateful for the friendships that came from that experience and for one especially beautiful and enduring friendship that continues on way past the fellowship's end date. *Grazie.*

I am also indebted to the "little decolonial reading group" of which I am a part, alongside fellow scholars Drs. Judith Krauss and Andrea Jimenez. Our discussions remain imprinted on my ever-evolving view of the world.

I am grateful also to the brilliant students at the University of St. Andrews for questioning my ideas in the classroom and for productively engaging with them in challenging ways. I am thankful for the freedom afforded me by the School of International Relations to pursue my own intellectual interests and to engage with them meaningfully even as they stray further and further away from the central concerns of the discipline. I am thankful for Professor Ali Watson's support in the closing phases of this project and to my colleagues at the Third Generation Project and at the Centre for Global Law and Governance, both research centers at the University of St. Andrews. Their engagement with my ideas and writing made this book markedly stronger and clearer. I am also thankful to the cohort of new staff members with whom I joined the University of St. Andrews in 2020, who, in the midst of the global coronavirus pandemic, managed to show warmth and collegiality across distance and repeated quarantines. I am especially thankful to fellow new starter Dr. Roxani Krystalli for being a steady supporter and critical engager with my work, and to Drs. Michael Simpson and Ifesinachi Okafor-Yarwood from the School of Geography and Sustainable Development for their generous collegiality, friendship, and support.

I am thankful to members of the Political Ecology reading groups at the Universities of Manchester, Sheffield, and Lancaster in the United Kingdom for providing me with feedback on earlier drafts of this book, and to those at the University of Cambridge for their willingness to engage with this project. Many thanks to Drs. James Fraser, Benjamin Neimark, and others at the University of Lancaster for their support, and to colleagues at Wageningen University in the Netherlands for continuously producing pathbreaking scholarship that stimulates my thinking.

I also thank the anonymous reviewers of this book and other related published works that helped me to flesh out the ideas that constitute the larger intervention made here. Their challenges, feedback, and critical engagement helped me to grow more confident in the realization that I did indeed have something to say.

I am grateful to my friends from home for their support across borders and communication channels, some of whom warrant special mention. I am thankful to Eutinde Madonna Allen, Derryann Edinboro, Christopher Ferrell, Roman Harris, and last but not least, Leon Belony, my dearest friend and lay reader.

I am eternally indebted to Malte Istel for his unwavering belief in me. I wish everyone would be fortunate enough to benefit from such resolute support.

I am immensely indebted to all the scholars on whose work and ideas I draw in this book, some of whom populate the reference list and some of whom remain unacknowledged because they indirectly inform my ideas. I am especially grateful to those scholars of Caribbean origin and background, on whose brave ideas, research, and scholarship I am able to rely in spaces hostile to my presence. Their scholarship is testimony to the fact that knowledge is not the exclusive domain of the privileged.

I reserve my final expression of gratitude to you, dear readers, for taking the time to engage with my ideas, for taking them forward and/or pushing them back in areas I might have misunderstood. *Thank you.* All errors, naturally, remain my own.

Introduction

Some three hours' drive from the center of Paramaribo, Suriname, along a single paved road through the Amazon rainforest, I sat in one of the satellite maroon communities that form Brownsweg, with two Dutch translators in a small wooden building. The building was the domicile of the community's captain. The home had a smattering of things. It was a mixture of modernity and tradition. On the way in, we saw maroon women sitting outside in a communal yard, topless and wearing the patterned print that characterizes their traditional dress, while tending to food in big bowls at their feet. Inside sat the community's captain wearing jeans and a T-shirt emblazoned with the letters *NYC*. He was seated on a white plastic chair with fragments of other places and times surrounding him, including a broken mixer, gas cylinders, and empty bottles of alcohol.

During our interview, the maroon captain explained that he makes his living by mining for gold beneath the forest floor. While expressing an interest in the forests being protected "for the future" and attributing that responsibility to the state government of Suriname, he noted that he himself, in his capacity as gold miner, causes trees to be cut down. In his expression of these conflicting interests and attributions of responsibility, the somewhat quotidian contradiction underlying the maroon village captain's

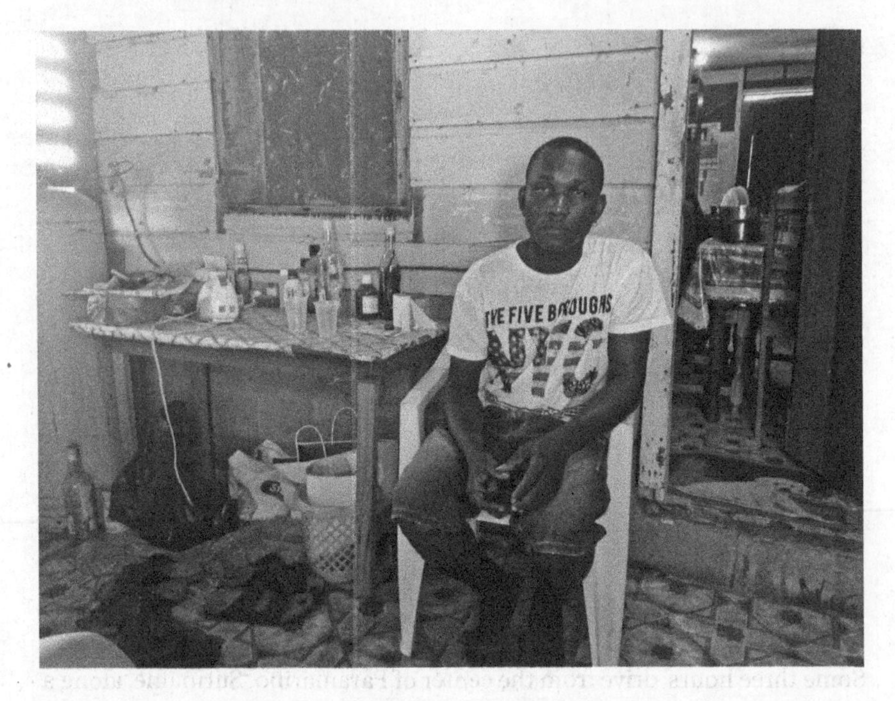

Figure 1. Maroon captain, Brownsweg, Suriname (Collins, 2014).

responses to my questions became clear. While contradictions of this sort can be easily dismissed as features of simply being human, his response still came as a bit of a surprise to me, given my awareness of the strong historical, cultural, and often spiritual connections of maroon communities to the lands they inhabit. Like the indigenous communities that also reside in the forests of Suriname, maroon communities practice traditional communal livelihood strategies to sustain themselves. They rely on community self-governing structures and cultural and spiritual practices that are closely tied to the forests and to their environments.[1]

These Amazon forests, stretching across Guyana, Suriname, and the wider Guiana Shield, have witnessed the entirety of the colonial and postcolonial encounter. Over centuries, the forests of the Guiana Shield provided a place of refuge to people who fled the coast to form the spatial outskirts of colonial centers in northern South America. During this time, the forests were themselves partitioned, regrouped, uprooted, and used as shelter from the whirlwind of capital-generating activity taking place on the coasts of the

Shield.[2] In the current era of rapidly accelerating climate change, the significance of these forests came to be even further enhanced as the international community began to recognize their sustainable management and conservation as crucial to global, ecological, and human well-being, with effects far beyond the boundaries of the Shield itself.[3] These forests comprise one of the last intact forest ecosystems on Earth and form a region of strategic import for supporting the well-being of the wider Amazon ecosystem.[4] Hence, the international community sought to financially incentivize their conservation through the United Nations–sanctioned Reducing Emissions from Deforestation and forest Degradation (REDD+) initiative, which emerged in the late 2000s as a tool for managing the world's remaining tropical forests.

If REDD+ should have worked anywhere, it should have been here. Avoiding deforestation should be relatively easy in Guyana and Suriname, two countries with some of the highest rates of forest cover and lowest rates of deforestation in the world. These features suggest a tradition of leaving the forests standing whether intentionally or by neglect. Yet, REDD+ in Guyana and Suriname has been fraught with challenges. The majority of the sparse population resides on the historically forest-denuded South American coasts of both Guyana and Suriname, leaving the inland forests primarily the domain of comparatively small numbers of gold miners, indigenous and maroon peoples. However, while tropical, forested countries in the Guiana Shield and around the world got ready to participate in REDD+, deforestation levels in Guyana and Suriname persisted or even increased. Although deforestation rates remained relatively low,[5] small-scale gold mining, the main driver of deforestation, continued to increase in its spread and intensity.

Some onlookers have since attributed REDD+'s global intractability and failings to its inability to overcome entrenched and powerful deforesting interests in participating countries,[6] and to the fact that the international cap-and-trade system for managing carbon, of which REDD+ was intended to be a part, never fully emerged. This situation led to an absence of funding for REDD+ activities.[7] Some onlookers also attributed these failings to REDD+'s reinterpretation within international finance practices and national policies in ways that are synonymous with largely ineffective conservation efforts.[8] Other more critical voices pointed to over a decade of REDD+ preparation and a paucity of results in order to support

Figure 2. Deforesting effects of gold mining in Guyana (photo credit: Malte Istel, 2022).

their declaration of REDD+ as a global failure, just another conservation fad that takes resources away from avenues that could otherwise arrive at meaningful change.[9] Some in this last group attribute this failure to broader, economic system dynamics such as the inability of capitalism itself to generate funds for conservation.

While some of these arguments surely have merit, in this book I take a different approach. Instead, I show how REDD+ is challenged not only by the structural failings of the global economic system it was constructed to suit, but also by its proponents' inattentiveness to colonial histories that have positioned the forests as places of refuge and resistance by people fleeing exploitation on the coasts. Hence, the entrenched and powerful economic interests pointed out by the first group of REDD+ onlookers may very well be colonially rooted ones.

Together, these chapters capture the histories of the forested places that became the independent states of Guyana and Suriname. They show how the forests, formerly a place of refuge from the exploitative machinations of capitalism, can now scarcely provide such a reprieve. Therefore, any attempt to decolonize forest use practices in states shaped almost entirely by the colonial encounter depends on the following combination of interventions. First, decolonization requires less deference to the sovereign state in questions of environmental governance. Second, it necessitates the removal of the market from its increasingly central position as arbiter of environmental and social affairs.[10] Third, it involves the undisciplining of the racialized subjects of colonial governance. Finally, it demands that those ethics and ways of being in the world that are associated with precolonial and non-Eurocentric knowledge traditions be amplified and taken seriously.

In developing these arguments, this book examines how recent international efforts to govern and conserve forests are being stymied by colonial histories, especially in forests shaped over centuries into places of refuge. In so doing, it imagines decolonial futures in Guyana, Suriname, the wider Amazon basin, and beyond.

WHY DECOLONIZE?

> Decolonization, which sets out to change the order of the world, is, obviously, a program of complete disorder. . . . Decolonization, as we know, is a historical process: that is to say it cannot be understood, it cannot become intelligible nor clear to itself except in the exact measure that we can discern the movements which give it historical form and content.[11]

"Why decolonize?" A pivotal question. Provided that a satisfactory answer can be put forward, the inquisitive mind quickly wonders, "But *how* would one even attempt such a thing?" This book puts forward an answer. My argument hinges, first and foremost, on the recognition that histories of colonization function as a structural condition that must be recognized and dismantled through *de*colonization if environmental governance is to successfully counter some of the world's most urgent environmental challenges.[12]

In the interest of limiting the progress of one such challenge—that of human-induced climate change—REDD+ rose to become the biggest, most ambitious global plan to arrest the deforestation of tropical forests around the world. It proceeded from the premise that tropical forests sequester significant amounts of carbon from the atmosphere as they grow. Carbon dioxide, one of the harmful gases driving climate change, is then kept trapped in the mass of the trees. As forests and surrounding vegetation decay and degrade, carbon dioxide is released back into the atmosphere. Hence, in the interest of keeping sequestered carbon dioxide locked away in the forests, REDD+ aims to provide financial incentives to forest managers, users, and owners to dissuade them from engaging in deforesting activities.[13] Put simply, REDD+ aims to stack the decision-making process on how to use forests against those decisions that result in deforestation. Since tropical forests hold more carbon than temperate or boreal ones, REDD+ is being rolled out in tropical states that still have significant tropical forest cover. However, these states, through little to no coincidence, tend to be those still strongly defined by their colonial pasts.

Consequently, the increasingly urgent need to address climate change by reducing carbon dioxide emissions brought with it a growing demand for the "services" provided by the forests of Guyana and Suriname in situ through states' participation in REDD+. Promoted as being able to make forests worth more alive than dead, REDD+ was championed by its proponents both within and outside of Guyana and Suriname as a tool for conserving the world's tropical forests, with the government of Guyana being one of its earliest proponents. Much like other market-based conservation mechanisms around the globe, REDD+ was framed as a "win-win" solution that would allow both countries to develop economically while conserving their forests for the greater global environmental good.[14] It was depicted as a market instrument imbued with a certain neutrality that allows it to sidestep political and social considerations around forest use practices, to bring about reductions in deforestation.[15] In the international arena, it was promoted by a variety of actors, including developing countries with high forest covers and international development banks, as a shortcut of sorts for reaching the "low-hanging fruit" of the climate mitigation game.[16]

In this book, I show how preparation for the eventual implementation of REDD+ in Guyana and Suriname made visible the structuring tendencies of the continued colonial present, thus demonstrating the need to

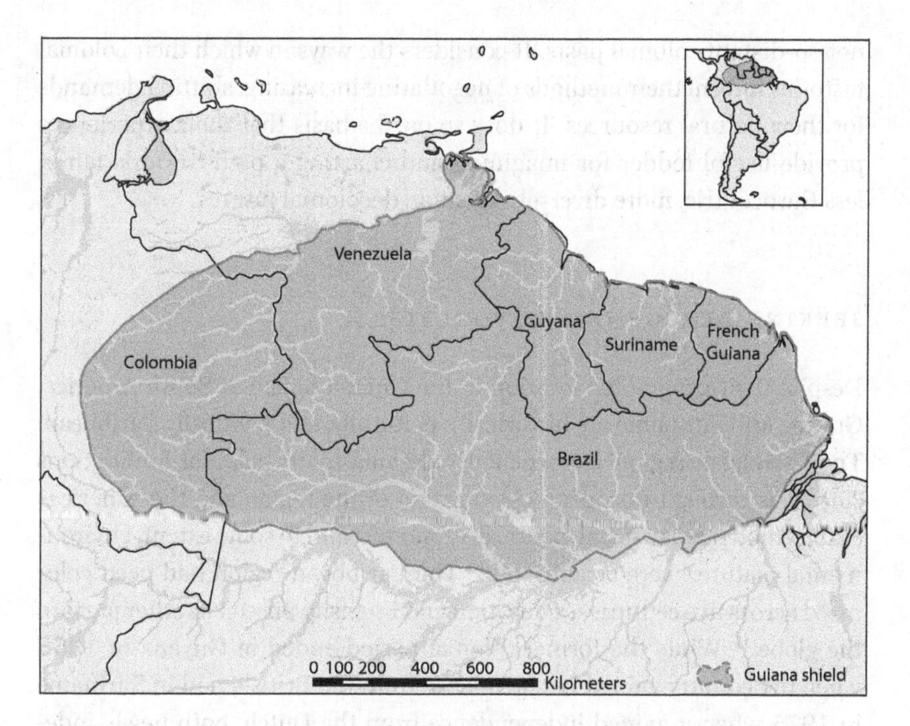

Figure 3. Map of the Guiana Shield (image credit: Oronde Drakes).

decolonize environmental governance. Decolonizing forest governance depends first and foremost on acknowledging how REDD+ is positioned within the histories and present-day social, economic, and political conditions of the jurisdictions within which it is implemented. Since these two neighboring countries are the only independent states, hence REDD+ participants, that are situated entirely within the Guiana Shield, their simultaneous exploration allows me to reflect on processes of state formation and demarcation as significant relics of colonialism with material consequences on modern-day environmental governance.[17] Here, it warrants mention that the forests of French Guiana, the other territory that lies completely within the Shield, are accounted for in the carbon management and accounting practices of the Republic of France. French Guiana remains French territory, having never gained independence from its colonizer. Still, this book recognizes Guyana and Suriname as independent, forested states in the Guiana Shield that were formed almost entirely through their not-so-distant colonial pasts. It considers the ways in which their colonial histories inform their methods of negotiating increasing, shifting demands for their natural resources. It does so on the basis that their experiences provide useful fodder for imagining and charting a path towards fairer, less Eurocentric, more diversely agential, decolonial futures.

SEEKING REFUGE IN THE FORESTS

Despite their geographic location in the Guiana Shield of South America, Guyana and Suriname are historically, culturally, and politically Caribbean. The Caribbean region is particularly defined by its colonial foundations since it is through European colonization of the region and the Americas that its modern-day social, political, economic, and to some extent, environmental features were established.[18] The Caribbean region had been colonized across five centuries, a duration and intensity unseen in other parts of the globe.[19] While the formal colonial period ended in Guyana in 1966 when the country gained independence from the British, and in Suriname in 1975 when it gained independence from the Dutch, both newly independent countries continued to be caught up in relations of dependency with the wider world in their post-independence periods.[20] In particular, their forest and mineral resources continued to be used as sources of raw materials exported for refinement and consumption elsewhere, with little economic benefit redounding to their origin countries.[21] Maroons, as communities and ethnic categories, were brought into being through and in resistance to these specific colonial histories, people, and natures.

Between the late 1500s and early 1600s, European explorers confronted diverse groups of people, who are now classed as indigenous to the Guiana Shield. Over time, the Europeans went on to establish colonies on the coast.[22] The centuries that followed this initial encounter saw a steady stream of initially indigenous people retreating to the forests and seeking refuge there from the capital-accumulating activities taking place on the coast. Over time, escaped enslaved Africans came to join them in setting up domiciles in the forests. These Afro-descended communities, eventually categorized as maroons in Suriname, had their migratory origins in the transatlantic slave trade through which millions of Africans, between the years 1501 and 1867,[23] were forcibly transported to the Americas to labor in colonies on plantations controlled by colonizers from Western Europe.[24] Maroons were able to live largely independently in the forests and to govern and sustain themselves and their resources,[25,26] though not without some occasional conflict with indigenous communities and the colonizers, who often collaborated to limit activities in which the maroons engaged, such as raids they conducted on the plantations to free more slaves.[27] In Guyana, the outcome was different. As enslaved Africans fled the plantations, they were systematically rounded up and returned to plantation owners by those indigenous communities that had been enlisted by the colonizers for this task, limiting the formation of officially recognized maroon communities there.[28,29]

After the close of slavery in the nineteenth century, Suriname remained a colony of the Netherlands until 1975.[30] During that period, the maroon communities of Brownsweg came to be categorized as transmigratory as a consequence of development paths pursued by the Dutch colonial government.[31] Brownsweg's residents were forcibly relocated there in the 1960s when the Dutch were completing construction of Suriname's Afobaka hydropower dam to provide low-cost electricity to fuel the then-booming bauxite industry. These communities, comprising approximately six thousand maroons, were forced to relocate. Those that refused saw their communities flooded. They lost infrastructure, land, grounds of spiritual significance, and even their lives as the dam's reservoir was constructed. Some of these communities were then relocated to Brownsweg. In the new, unfamiliar environment, some of the relocated maroons, among them the parents of the maroon captain with whose interview I opened this book, turned to small-scale gold mining to earn an income. While themselves a casualty of the demand for low-cost energy for mineral production, some community members eventually also turned towards mineral extraction as a means of sustenance.

All in all, for indigenous communities, now known collectively as Amerindians in the region, refuge meant retreat from the coastlands through which they had once freely roamed, hunted, and sustained themselves. For escaping enslaved Africans, this movement formed part of the resistance that saw maroon communities developing in Suriname's rainforests as they learned to live in these areas from the indigenous people

Figure 4. View of flooded-out area and dead trees from where maroon communities now comprising Brownsweg were displaced, seen from the Brownsberg nature reserve.

they encountered, and deployed techniques they brought with them from Africa. During the colonial period, seeking refuge in or retreating to the forests was decidedly anti-colonial.

REDD+, in its formulation as a market-based initiative for altering deforesting behaviors, such as gold mining, through the provision of financial incentives, attempts to sidestep the histories through which current forest use practices became entrenched. In so doing, REDD+'s implementation on the ground is inevitably confronted by modern expressions of the historical processes of colonialism, racialization, and capitalism-fueled extraction that shaped the forests of the Guiana Shield and the behavior of people living within and using them over centuries.[32] By offering incentives for avoiding deforesting behavior, proponents of REDD+ imagine those who engage in deforesting activities as responsive subjects of market-based governance,[33] sidelining the complex, seemingly incoherent behavior of those residing in and managing the forests, such as that of the maroon captain with whom I opened this book.

Even further, in its focus on market transactions for quickly stemming the tide of deforesting behaviors, REDD+ reduces the significance of the colonial histories through which people became threats to the forests. Instead, REDD+ repackages the complexity of deforesting behavior and its outcomes within the logic of the market. However, it is precisely through the exploitative, historical events from which maroons had initially fled, as this book will show, that market logic began its ascension to centrality in the management of human-environmental relations in Guyana, Suriname, and elsewhere. It is precisely in these events that climate change has roots.[34]

Now, as the international community grapples with the debilitating ongoing and predicted effects of the changing climate, greater attempts to conserve the forests are being made. The forests, recast as cheap, effective allies in the fight against climate change, are again providing refuge, but this time explicitly to the wider international community. In a repeat of past events, the people left behind in previous iterations of modernity's forward thrusts are being held responsible for remedying its failings.

COLONIALISM AND THE COLONIALITY OF POWER

Colonialism has been defined in a variety of ways.[35] As the system emergent from processes of colonization, colonialism generally refers to the enforced control of space, land, bodies, and nature by an outsider who may become localized. Colonialism is made possible by extracting and relocating resources to the colonizers' motherland primarily for the benefit of their community; or by appropriating space and resources for their exclusive use locally.[36] Colonialism in spatial view of the environment can be understood through its settler ramifications, where significant numbers of people relocated and settled in new territories, or through nonsettler processes of establishing overseas dependencies.[37] While both these forms of colonization involve the act of taking over of space, either through control undertaken remotely or through permanent resettlement in colonized spaces, spatial conflicts over access to land are more immediate, visible, and physically palpable in settler forms of colonialism.

In this book, I pay attention to nonsettler colonialism that molded the Caribbean region,[38] and I rethink decolonization in a way that necessitates engagement with the transatlantic slave trade that, in no small part, shaped the region's (mis)fortunes.[39] I draw heavily on the work of French philosopher Michel Foucault to identify and name the strategies of governance I see as functioning in support of colonization and supporting its resultant colonial structures. Those very strategies, I argue, are the ones that should be used in efforts to *de*colonize environmental governance. As a result, I frequently use the term *(de)colonization* to demonstrate that those strategies that supported European colonization of the Americas are precisely the ones that should be used as entry points into discussions of how decolonization can be pursued.[40] Hence, this book offers to the burgeoning literature on decolonization in relation to the environment an interrogation of nonsettler colonialism that is bound up with those exploitative and state-forming events that fueled capitalism's rise.[41,42]

The outward spread of colonialism's structuring effects from colonized places to the rest of the world was compellingly conceptualized within Anibal Quijano's description of the coloniality of power.[43,44] Quijano put a name to colonialism's continued structural and societal effects past the close of its formal end date and outside the site of its pursuit. He recognized the colonization of the Americas as foundational to today's global power constellations, which underpin the very structure of capitalist modernity.[45] In this sense, modernity, partly a consequence of capital accumulated through the colonization of the Americas, is eternally inflected by histories of colonial domination—histories that continue to structure the relationship between the West, as the victors of colonialism, and the rest.[46] Hence, colonialism and modernity, along with the environmental challenges they bring, are inextricably linked. Therefore, in the words of Walsh and Mignolo, "to end coloniality it is necessary to end fictions of modernity."[47]

Coloniality is therefore a concept that challenges commonplace interpretations of colonialism as being restricted by its temporal span ending (formally) at independence, and by its geographic span being limited to colonized spaces. In like manner, the colonial histories that defined the emergence of Guyana and Suriname continue to structure societies and power relations both within and without the forests that have borne witness to these events. Colonial histories understood in this way can be imagined in simplified illustrative form as concentric. In the shared center of my understanding of coloniality in Guyana and Suriname lie the forests where indigenous and maroon communities continue to embody colonial histories of state creation, living on land apportioned or allowed them by the colonizers in the ways in which they have grown accustomed. In the first outer circle lie the coasts of Guyana and Suriname where different groups of people brought in waves over the centuries to form the labor force continue to be spatially distributed, politically affiliated, and related to the creation of capital along the fault lines established during colonialism. In the second outer circle lie the substantial revenues generated from the exploitative, colonial activities that fueled the rise of capitalism in colonial centers in Western Europe.[48] Finally, reaching even further still, lie the capitalist structures increasingly encircling the globe.[49]

Coloniality is also made visible through race.[50] In Guyana and Suriname, race is the outcome of processes of racialization through which people were stripped of the ethnic categorizations, names, and groupings to which they had previously ascribed, and were sorted instead according to newly inscribed bodily markers that fixed their relationship to land, labor, and capital.[51] Thinking again concentrically, as colonialism went on to fuel the rise of capitalism in Europe,[52] so too did conceptions of race in the Americas fuel global conceptions of race as constitutive of and inextricable from capitalist modernity.[53] So, it came to be, through colonial histories of conquest and exploitation in the Americas, that race became a biological marker of those who were conquered and those who conquered.[54] Over time, these relations were naturalized to explain not just external differences between the conquered and the conqueror, but also their mental and cultural differences. The category of whiteness became associated with that which is superior, and the category of the nonwhite with that which is inferior. Colonialism marked these processes of racialization as political projects over time and space.[55]

The coloniality of power makes clear the racialized continuities between the emergence of the capitalist economic system and colonialism.[56] In the Americas, economic processes associated with the accumulation of capital became the axis around which other social relations were organized.[57] The exclusive control of the circulation of resources produced in the Americas by those categorized as "Whites,"[58] along with the concentration of the commodification of the labor force for "White" workers, meant that capital "as a specific social relation, was concentrated in the geographic region that then received the name of Europe."[59] In this context, "Europe or, more specifically, Western Europe emerged as a new historical entity and identity and as the central place of the new pattern of world-Eurocentered colonial/modern capitalist power."[60] Through colonial dynamics, histories, and interactions, race arose to become a "complex assemblage of phenotypes and environments rearranged by colonialism and capitalism . . . [and] the material and mental division of bodies into groups according to shifting criteria."[61]

As anthropologist Patrick Wolfe observed, while race is indeed socially constructed, it is more importantly a site-specific "trace of history" through which "colonised populations continue to be racialised in specific ways that mark out and reproduce the unequal relationships into which Europeans have co-opted these populations."[62,63] Therefore, the interests of oppressed groups often run counter to each other because the groups were exploited and integrated into the colonial enterprise in different ways. Race is therefore "colonialism speaking, in idioms whose diversity reflects the variety of unequal relationships into which Europeans have co-opted conquered populations."[64] These social constructions do not merely operate "on a naturally present set of bodily attributes that are already given prior to history."[65] Instead, "racial identities are constructed in and through the very process of their enactment."[66]

Often overlooked in these discussions on the social construction of race, however, is the key role played by the environment.[67] The understanding of race as constructed according to and through external criteria is usually reflective of the relationship between the human body being racialized and something or someone external to him or her. As geographer Juanita Sundberg notes, projects of political economies, colonization, and/or natural resource allocation come together to influence how race takes shape.[68] In other words, the environment is not merely a passive actor in the background, but is, in fact, actively relationally involved. Therefore, colonialism produces not just localized and globalized conceptions of race, but also racialized relations to the environment that are often mediated by labor.[69] Anthropologist Melisa Johnson referred to these racialized relations with the environment as "socionatural becomings" as she demonstrated how the group identities of some human beings emerged through entanglements with nature in the Caribbean.[70] Her work on Afro-descended communities in Belize presents one of the few openings into how different groups of nonwhite people negotiate the dynamics of climate change and race in view of the Anthropocene, which is seen by some as the current geological epoch through which human beings are the dominant force in shaping the environment and the climate.

All in all, during colonialism, people found in and brought to Guyana and Suriname were categorized and racialized by European colonizers according to the colonizers' interpretation of their value. In their current post-independent condition, these dynamics have shifted somewhat since the colonizer, categorized and represented as white, is no longer physically present in large numbers with colonialist backing on the ground.[71] Race, however, remains instrumental in how people relate to capital, to forests, and as a corollary, to REDD+.

THE "HOW" OF DECOLONIZING ENVIRONMENTAL GOVERNANCE

Guyana's independence from the British was influenced by external powers from its very outset.[72] On the other hand, Suriname's independence from the Netherlands saw the emergence of a paternalist relationship with its former colonial master and marginal engagement with the international community. The independent governments that came to govern Guyana and Suriname began to recast the forests, historically a place of refuge from coastal capital-accumulating activities, as natural resources to be exploited to fuel development and modernization goals.

Maroon communities persisted throughout these shifts from colonialism to independence, after having been established on decidedly anti-colonial grounds. To them, the forests stood as the antithesis of capital-generating activities centered on the coasts. It was this pursuit of capital that had enslaved them, subordinated nature, created colonies, and consolidated all colonized land, including land claimed by indigenous and maroon communities, into independent states. Although the anti-colonial gestures of maroon communities were eventually subsumed within modernity's onward global march, they remained imprinted in the eventually independent Guyanese and Surinamese states.

Decolonization, however, holds modernity's Eurocentrism to account while making visible those ways of being that were marginalized by colonialism for centuries. It does not default to a romanticization of the past but advocates for a dismantling of the colonially structured present, or of what Malcom Ferdinand might refer to as colonial inhabitation.[73] Decolonization directly challenges the overrepresentation of Europe in knowledge systems and other post-independence ways of being that were dominant in formerly colonized states.[74] It requires the dismantling of colonial structures in a way that examines processes and histories of racialization manifested in the very existence of the Guyanese and Surinamese states since it was through colonialism, slavery, and indentureship, operating to varying degrees in service of capitalism, that these states were established.

(De)colonial Mentalities

In the chapters that follow, I chart a path towards decolonizing forest governance in countries almost completely defined by the colonial enterprise. I recognize that for centuries, a drive towards improvement for select groups of humans made life better for some. The bodies and energies of those less fortunate others came to be systematically used as fuel for satisfying this drive or were simply made disposable in its wake.[75] Yet, in recent times, arguably defined by the unfolding of the unintended consequence par excellence referred to as human-induced climate change,[76] capitalism charges onward to the detriment of the natural fabric that supports all life on Earth. Modernity and its linear progress narrative have brought humanity to the point where its unintended life-threatening consequences can no longer be viewed simply as external to its own existence and development. The climate-changing, biodiversity-erasing consequences of capitalist modernity can no longer be relegated to the category of its problems that can be addressed merely by recalibrating codes and algorithms in the hope that the problems can be simply fixed and capitalism can be sent chugging along its way. The scale of the environmental challenges faced by human societies and the increasingly perceptible impacts of human-induced climate change threaten life itself and further magnify those threats posed to those lives deemed disposable in modernity's rise.[77] Hence, the elephant that *is* the (modern) room must be reckoned with—the figurative elephant that obscures attempts to think outside of the figurative room, limiting efforts to question the linearity of modernity's progress and the capital-driven, life-selective nature of its nurture.[78] It is only when we consider that the modern is intricately bound up with questions of the colonial that we begin to appreciate the urgency of the impetus to decolonize.[79] Attempts to decolonize or to grapple with how decolonization can be approached provide cracks in the shiny surface of modernity that allow the inquisitive mind to peer outside, to reevaluate what is important, and to imagine other ways of being. But anti-colonial,[80] now-decolonial,[81] gestures have always been with us. Their marks remain if one knows where to look.

So how can decolonizing environmental governance be imagined in countries that are almost entirely defined through their colonial experience? Necessarily radical, this imagination recognizes some of the previous "anti" and now "de" colonial gestures that have always been present throughout the colonial and post-independence period in critically engaging with the colonial present. Forest governance in support of environmental conservation outcomes provides a useful entry point since conservation represents a mediating force between pressures for continued capitalist accumulation and efforts to maintain environmental function and well-being in the face of these increasing pressures.[82] Hence, environmental conservation cannot be decolonized as an entity separate from the broader societal dynamics in which it operates.[83] Recognizing environmental conservation as a mediating force requires that decolonial efforts engage not just with the strategies, norms, and logics that characterize conservation,[84] but with the broader dynamics that bind it to particular colonial pathways, structural conditions, and effects.[85]

Naturally, paying attention to the broader dynamics that influence forest governance and its environmental conservation goals significantly broadens the scope of this book. It forces me to move from the relatively narrow confines of environmental conservation projects that are usually premised on knowledge traditions that view the environment as separate from its human stewards, to societal-environment relations that are perceptible at the local level but also reflective of global considerations.[86] It also requires that I pay attention to the roughly five hundred years of history that influence land and forest use traditions in Guyana and Suriname. The scale of this endeavor is considerable. It demands that I draw upon a coherent conceptual framework to identify, label, and trace the logics of governing the environment and its users of a particular place over time.

European in origin,[87] poststructuralist Foucauldian insights prove particularly useful in supporting my analysis of colonial and post-independence governing strategies of the land that eventually became Guyana and Suriname. These governing strategies, or "govern-mentalities," provide the scaffold on which REDD+ and some other environmental conservation initiatives build.[88] Through what I recast as (de)colonial mentalities, I draw on Foucault's governmentality framework for labeling and examining the overlapping governing strategies employed in enacting forest governance over time. Together, these strategies capture the shifting logics of colonialism in operation, while allowing them to remain identifiable in these countries' transition past the formal end of the colonial period. In so doing, I view colonial governing mentalities as sites for decolonization, hence the use of the term *(de)colonization.* I assert that we can begin to imagine a path towards decolonization by first recognizing how people and their environments were governed throughout the colonial and post-colonial periods and by remaining aware of how nonsettler colonialism continues to be impactful in the structures of these societies even after colonialism's formal end.[89]

My neologism, *(de)colonial mentality,* therefore builds on Foucault's influential governmentality concept that seeks to highlight the "conduct of conduct" embodied in the multiple approaches to governance more broadly.[90,91,92] Governmentality is often deployed as a means of understanding how governments seek to achieve some form of improvement in their populations, manipulating the populations' behavior by drawing on different types of knowledge, judgments, and ways of doing things.[93] Governmentality takes multiple forms that can be distinguished through their approaches for achieving the desired behavioral change based on the sets of knowledge from which they are derived.[94] Multiple governmentalities provide us with insight into "how governing is accomplished in practical and technical terms."[95] Governmentality is understood as an art of influence through which diverse actors that are imbued with the ability to govern, even if just temporarily, work to shape outcomes and to guide the population within their remit in the desired direction.

Building on this perspective, the "multiple governmentality," or in this case "multiple (de)colonial mentality," approach provides a framework through which the different, overlapping governing strategies that operate in concert to achieve a particular outcome can be identified. These multiple (de)colonial mentalities include neoliberal,[96] truth, sovereign, and disciplinary forms,[97] each representing different mentalities of colonial governance. Neoliberal (de)colonial mentality "seeks merely to create external incentive structures within which individuals, understood as self-interested rational actors, can be motivated to exhibit appropriate behaviors through the manipulation of incentives."[98] The subject envisioned within this mode of governance is the ideal neoliberal *homo economicus* imagined as manageable and responsive to modifications artificially introduced in the environment.[99,100] *Homo economicus* is the imagined subject most amenable to the incentive-based governing logic of REDD+.

By contrast, sovereign (de)colonial mentality functions through the imposition of formal rules and regulations under the threat of force, often by independent state governments. Disciplinary (de)colonial mentality takes a less severe approach. Disciplinary methods of governing depend instead on encouraging the population being governed to internalize particular norms and values that should then lead them to behave in certain ways. Finally, truth (de)colonial mentality appeals to a "natural" order of things,[101] as demonstrated in religious texts and indigenous epistemologies and worldviews. Altogether, sovereign, neoliberal, and disciplinary (de)colonial mentalities operate through calculation and rationality, for example, through a reliance on censuses or statistics.[102] Truth (de)colonial mentality, on the other hand, is less rational in its orientation.[103] It functions instead according to "the truth of religious texts, of revelation, and of the order of the world."[104] I depict these graphically in table 1 for ease of reference.

These multiple, overlapping (de)colonial mentalities are each fleshed out individually in the chapters that follow. I use them to identify the overlapping scaffold of governing logics that supported the creation and management of colonized places and that continue to support forest governance and environmental conservation strategies today. These (de)colonial

Table 1 Multiple (De)colonial Mentalities

Mentality	*Description*
Discipline	Governance by encouraging the internalization of norms and values
Sovereign	Governance through the top-down creation and enforcement of regulations
Neoliberal	Governance by manipulating external incentive structures
Truth	Governance in accordance with a particular conception of the nature and order of the universe

NOTE: Adapted from Fletcher, "Neoliberal Environmentality," 178.

mentalities provide the tools and language necessary for tracing the contours of shifts, continuities, and disruptions in colonial forest governance over time. They further demonstrate how REDD+ builds on and is challenged by the legacy of these governing interventions. Most importantly, however, they provide tools for outlining how colonial governing strategies can be tackled and undone.

Methods and Chapter Breakdown

Efforts to decolonize my thinking, and to dismantle the colonial structure supporting modernity that defines the West,[105] would also require a particular dismantling of myself. Though racialized as black or Afro descended, I am not African. Though a native English speaker, I am not English. Though categorized by passport as Guyanese, I think through the British colonial frameworks that dominated the early stages of my education. I think through the archaic education and cultural systems that taught me, either through inference or explicit expression, that despite the everyday occurrence of 30-degree-Celsius weather, my bare arms are unladylike and my exposed knees, un-stately.[106] I was taught to pray through the religion that upended those expressions through which some of my ancestors prayed, and to identify with the practice of faith whose practitioners looked on as my ancestors were captured, raped, and tortured on their way to the land that is now etched on the document that determines my mobility. I grew up in land apportioned to me by the colonizers, land through which the numerous communities indigenous to that land had roamed before the forceful and contractually bonded relocation of my African, Indian, and Chinese and even Portuguese ancestors by (other) European colonizers forced them from that land. An examination of colonial structures and the ways of upending them requires an examination of me, a parsing of the different ethnicities brought to the shores of South America to labor in varying gradients of exploitation; ethnicities that continue to argue amongst themselves for the ruins that remain now that the colonizer in fleshy form has receded. In other words, I am the product of a colonized, racialized, development-infused environment that is oriented towards images of progress modeled on the West. This book forms part of my process of unlearning some of its principles.

The arguments I make in this book are developed in full view of my own relationship to decolonization and environmental governance. In my field research, I undertook participant observation in REDD+ preparation activities and events mostly from the position of a "local," since I, the author, am a national of Guyana. This familiarity extended to neighboring Suriname, where even though I was recognized as not Surinamese, due at least in part to my inability to speak the colonial language of Dutch, the local language of Sranantongo, or any of the many community-based languages spoken there, I was received in friendly ways that appeared to recognize the particular brand of Caribbean-ness and South American–ness shared by Guyana and Suriname. The overlapping cultural and migratory patterns of Guyana and Suriname permitted me easy passage through the Surinamese society, and a shared "blackness" informed and supported my presence in Suriname's forested areas, although I received numerous curious looks from schoolchildren while in maroon communities.

In addition to having my life nurtured and spent mostly in this region, I support the arguments made in this book through my undertaking, in 2014 during the early preparatory phases of REDD+, of a multisited ethnography of the initiative in Guyana and Suriname. During this early phase of REDD+ readiness preparation, I was able to gain insight into the expectations being made of the initiative as it negotiated the intersection between its development as an international policy and its eventual implementation on the ground. I travelled to four different deforesting hotspots in the two countries and spoke to representatives of the different stakeholder groups that were identified in national REDD+ preparation documents as instrumental for REDD+ outcomes. These travels were facilitated in part through my engagement in two internships with international REDD+ implementing organizations in Guyana and Suriname within United Nations Development Programme (UNDP) offices. I also interviewed over sixty persons identified as stakeholders of REDD+ and analyzed dozens of policy documents. Despite my every attempt to be thorough, the experiences presented here are partial. They are inflected by the manner in which I was received and perceived by the different actors related to REDD+ who, based on the ease with which I could be categorized in these locations, assumed my affiliations, political and otherwise, based on my speech, gender, skin color, and institutional affiliations, both local and overseas, academic and professional.

In the pages that follow, I explore the implementation of and resistance to REDD+ in Guyana and Suriname by connecting their colonial histories to the racialized environments, politics, and forest use practices that currently characterize both countries. I show how colonialism continues to influence Guyanese and Surinamese society in a manner that weakens ongoing international policy efforts aimed at conserving forests and governing climate change.[107] I do this while arguing that climate change and development trajectories are gradually eroding the historical ability of the forests to function as a place of refuge. I show how the global circulation of capital in which the Guyanese and Surinamese states have always been embedded supports human-environment relationships organized according to racial difference. REDD+ joins this global circulation of capital in ensuring that the potential for withdrawal from its accumulating regime is further eroded. There is no longer an escape.

The first chapter introduces what I refer to as the climate change and development nexus within which I situate REDD+'s pursuit. This nexus contextualizes the emergence of the REDD+ mechanism in Guyana and Suriname and identifies some reasons for its adoption. The second chapter presents the first of four steps I see as essential to decolonizing forest governance in the Amazonian Guiana Shield. This second chapter advocates for a figurative beheading of the sovereign by highlighting the sovereign power being resisted by maroon, indigenous, and other forest-dependent people.[108] This resistance is often directed at those actions and histories that violently disrupted the forest use traditions of indigenous people and that claimed the land currently known as Guyana and Suriname as European colonies. Further, it is directed at those actions that stratified societies and economic-environment relations racially and that continue to allow the now-independent governments to issue concessions in lands communities consider their own. These violent disruptions, forming the required first step for forest conservation and governance, are captured in the expressions of sovereign (de)colonial mentality I identify. There, I demonstrate why efforts to decolonize environmental governance depend on the recognition of the coloniality that is embodied by the Guyanese and Surinamese states.

The third chapter continues in the project of engaging with sovereign (de)colonial mentalities by advocating that markets be decentered from their de facto position as arbiter of social, political, and environmental policy. In it, I trace the pursuance of development models in Guyana and Suriname in the post-independence period by contrasting the eventual embrace of market principles in Guyana with the reclusion and exclusion from the international community of Suriname. I describe how the eventual dependence on market-led development set the tone for market-based conservation by instilling, in Guyana especially, a reliance on the market as the central organizing concept around which development was envisioned and made possible.

The fourth chapter identifies the undisciplined subjects of REDD+ governance. In this chapter, I outline the resistance expressed by people on the ground in both countries to the dictates set out by REDD+ proponents. I draw on disciplinary (de)colonial mentality, which seeks to foster the internalization of certain norms and values into the population being governed in the aim of shaping their behavior, to describe how actors related to REDD+ challenge its associated reliance on the market as an arbiter of social relations. The fifth chapter then advocates for market discipline to be countered with truths, as deeply rooted, often cultural or religious viewpoints on the order of the world. In it, I show how forest-dependent people are integrated into REDD+ through disciplinary governing methods that further complicate their racialized subjectivities. There, I trace how racialized subjectivities embedded in the colonial period and evident around the connected issues of gold mining and land rights respond to REDD+.

Altogether, these chapters provide a framework for conceptualizing and pursuing decolonial futures in places created almost entirely through colonialism. In each chapter, I begin to imagine a historically informed, decolonized practice of environmental governance that is more contingent on social, political, and economic arrangements than an outcome-driven intervention into these arrangements. This decolonial practice would be decentralized, without being dependent on state resources or its organization. It would not be capitalist and market centered,[109] and it would prioritize instead other ways of adjudicating between competing demands on nature. It would take seriously the questions, provocations, and manifestations of resistance expressed by those with a vested interest in its processes, without insisting on forcefully subsuming their views under the rubric of the desired outcome. Finally, it would draw on the knowledge, practices, and ways of life of these groups of people who instead of dominating nature, learned to live with and in it.

I note that during the final phases of writing this book, Guyana and Suriname, in the face of global environmental challenges that threaten the very survival of their human and nonhuman endowments, found significant amounts of oil in their territorial waters, the magnitude and newness of which are likely to aggravate the development, climate change, and racialized social and environmental dynamics this book describes. On 20 December 2019, the then president of Guyana, David Granger, declared that date "National Petroleum Day" to commemorate Guyana's official commencement of oil production. Days after this declaration, ExxonMobil announced its fifteenth oil discovery in Guyana's territorial waters,[110] which will add to six billion estimated barrels already found there. These discoveries stimulated an intensification of offshore oil exploration activity in Suriname, where much smaller amounts of coastal oil reserves have been extracted and refined since 1982. Large oil deposits were also subsequently found offshore in Suriname.[111] Both countries' populations cheered in anticipation of the wealth that would come to their shores and that should ideally allow them to modernize and improve their quality of life.

In view of these and other events, I join through this book a growing number of calls for the active unlearning of Eurocentric development models in the race against environmental disaster by investigating localized potential for decolonization.[112] At the same time, I root these arguments of processes of racialization in the relationship of different ethnic groups to the natural environment. Hence, the book reasserts the specificities of place-based histories in interrogating the site-unspecific nature of global climate change. It outlines how environmental governance and resistance to that governance evolved over time in providing a radical imaginary of how environmental governance can be decolonized.

1 Between a Mine and a Hard Place

Climate change is known to be "a messy mix-up of time scales" due to its characteristic absence of a clear and linear temporal correlation between a singular local cause and its effect.[1] Nevertheless, it refers broadly to those changes taking place in the Earth's climate due to the effects of global warming that result from the concentration of greenhouse gases.[2,3] Climate change places ecosystems, economies, and societies at grave risk, especially those that have done the least to contribute to its emergence.[4]

The global effort to address climate change relies partly upon environmental conservation initiatives to stave off its worst effects on human and nonhuman well-being. Environmental conservation tends to be enacted in those parts of the world that conservation practitioners, organizations, and other related actors deem natural and ecologically valuable enough to be conserved. Often, these areas are situated in places that have not yet been industrialized, in countries classified as the Global South. Actors in the industrialized North tend to favor this arrangement because conserving nature in the nonindustrialized Global South is often cheaper and does not require them to curtail lucrative industrial activities elsewhere.[5] For many onlookers, this global structural condition of the conservation movement is colonial in itself. It empowers wealthy, industrialized countries with moral arguments that support their interventions into the affairs of financially worse-off, nonindustrialized countries, while making insufficient changes in their own behaviors.[6]

However, this argument is simplistic. It overlooks the agency of those actors in the South who, much like their Northern counterparts, wholeheartedly embrace the market as an arbiter of environmental and social concerns. Further, deeming the conservation movement to be colonial because most of the funding comes from former colonial centers in the Global North has the often-perverse effect of recentering the North and bolstering its dominant position in discussions of how to address this challenge. This simplification keeps countries in the Global South from reckoning with how their pursuit of development paths that align with the global standard contributes to the "unintended by-product par excellence" of climate change.[7]

Without seeking to diminish the responsibility of industrialized countries for addressing their role in climate change, I recognize the ability of countries like Guyana and Suriname to chart their own path within the constraints imposed on them by the neoliberal, capitalist system. The staunch advocacy for REDD+ in the early days of its conception by Guyana's former president, Bharrat Jagdeo, exemplifies this ability. In fact, REDD+ was initially conceptualized and put forward to the international community by the Coalition for Rainforest Nations, a group of forested countries in the Global South. Therefore, while the primary source of funding for conservation organizations and initiatives around the world can indeed signal continued or intensifying colonial dynamics and relationships, it should not be the default, determining characteristic when qualifying these relationships as colonial.

Deforestation is a significant contributor to climate change.[8] Avoiding deforestation is therefore a pressing issue of global concern. However, this need to avoid deforestation in tropical countries of the Global South centers questions of forest ownership and use practices, which, when interrogated seriously and deeply enough, bring histories of colonization to the fore. Therefore, attempts to avoid deforestation in formerly colonized places make the need to decolonize environmental governance even more pressing. In developing this argument, this chapter highlights the development-related drivers of deforestation through which state governments exploit natural resources in the hope of improving the living standards of their populations. Further, the chapter highlights the externally driven drivers of deforestation through which goods are produced to satisfy the demands of the international market. Put simply, it shows that the threats to the forests of Guyana and Suriname do not come solely from without.

Overall, this chapter situates the progress made by Guyana and Suriname towards the implementation of the Reducing Emissions from Deforestation and forest Degradation (REDD+) initiative within a set of circumstances I refer to as the development and climate change nexus. Commencing with an exploration of the specific vulnerabilities to climate change confronting the Caribbean region, the chapter traces the varying threats to the sustainability of the forests of Guyana and Suriname that have roots in the pursuit of nebulous ideas of development. Hence, the chapter explores the relationship between development trajectories and market-based conservation to provide the necessary context for understanding the social and political challenges associated with managing and maintaining relatively low, but increasing, levels of deforestation in these two countries.

CLIMATE CHANGE AND THE CARIBBEAN

The Intergovernmental Panel on Climate Change (IPCC), an intergovernmental body of the United Nations that advises the international community on scientific and technological aspects of climate change, surmised that global temperatures are likely to increase between 1.4 and 5.8 degrees Celsius this century. The temperature increase depends, of course, on the success of human interventions in limiting the rise of greenhouse gases.[9] As the international community began to collectively coordinate its response to climate change, some Caribbean states began to coordinate their response through the region's integrating body called the Caribbean Community (CARICOM), in the hope of giving the region a bigger collective voice in the international arena. CARICOM came into being in 1973 in recognition of its members' shared experience of "Caribbean-ness," which derived in part from their colonial pasts, subjugation, and the initial forced creation of societies by external forces.

The effects of the changing climate, while likely to be dramatic around the globe, are particularly acute on the small islands and low-lying coastal states of the Caribbean region. These states are particularly vulnerable to coastal erosion and sea level rise.[10] Increasing human-induced pressures on coastal areas will likely multiply this vulnerability.[11] Flooding is anticipated to be a consistent threat "exacerbat[ing] inundation, storm surge, erosion and other coastal hazards, thus threatening vital infrastructure, settlements and facilities that support the livelihood of island communities."[12] These climatic effects have been increasingly evident in recent years as extreme weather events have had a devastating impact on the Caribbean region. For example, during the period of 1995 through 2000, the region experienced the highest level of hurricane activity, including tropical storms and flooding, amounting to significant financial losses. Delays in responding to climate change will further increase these costs.[13]

DEVELOPMENT AND DEFORESTATION

Deforesting activities in Guyana and Suriname stem from unsustainable forestry practices, foreign investment / large-scale infrastructure projects, and gold mining. Forestry accounts for a significant portion of deforestation and forest degradation in both states. Both countries have very stringent forest-monitoring guidelines that restrict where trees can be harvested, the way this should be done, the establishment of buffer zones, and no-go areas. These regulations prescribe how the forests should be utilized and allot penalties to code breakers. However, in both countries these guidelines are not being respected, especially by foreign companies accused of being responsible for some of the largest infractions. Even though international companies are not permitted to own timber concessions in Suriname, they have been able to buy the concessions of others operating there.[14]

The second significant source of deforestation in Guyana and Suriname involves infrastructure and foreign investment, especially related to road building. Foreign companies play a significant role in this industry, as exemplified by the role of Chinese companies being awarded infrastructure projects within Suriname, some of which were highly contentious and allegedly corrupt. These projects also threaten biodiversity in these two countries as forests are opened up and made increasingly accessible, leading to a rise in recreational hunting within the forests by both local and foreign actors. The drive for resources also comes from the growing appetite for resources in Asia. In Guyana, the Chinese have been strengthening their hold on mining and lumber concessions, with gold mining concessions forming the bigger threat since the law allows for forests to be clear-cut to access gold beneath the soil. One such problematic concession is that of the now-defunct Bai Shan Lin, a Chinese company that was reportedly clear-cutting large sections of the forests in 2014 through mining and lumber licenses awarded by the government. After allegations mounted that the company was not paying its share of taxes and was not respecting the rules and regulations of Guyana, its concessions were repossessed by the Guyana Forestry Commission in 2016. Representatives of an environmental nongovernmental organization had also expressed concern that, in both Guyana and Suriname, the selling off of natural resources to the Chinese was done without considering sustainability and the well-being of the forests.[15]

The international challenge of foreign concessions was highlighted by an adviser of the state government of Guyana who stated, in his interview with me:

> We don't have the critical mass in terms of population to develop this country, so currently, you have Brazilians here. You have Indonesians. You have Malaysians. You have Chinese. You have people from the Caribbean, Trinidadians, and so on, all of whom are utilizing Guyana and Guyana's resources, land, and whatever and its geopolitical, geostrategic position in a sense like an aircraft carrier for the rest of South America.[16]

The dominance of foreign nationals in extractive activity in the forests is especially perceptible around gold mining, which is dominated by Brazilians who brought to Guyana and Suriname the technologies that are most effective at extracting the largest amounts of gold possible from the soils. In retelling the story of the northward movement of Brazilian garimpeiros who have come to be strong actors in gold mining in Guyana and Suriname,[17] in terms of both their numbers and their influence on the mining industry, a leading figure in gold and diamond mining in Guyana, interviewed in 2014, explained:

> The Brazilian government at one time, as a social experiment, allowed people to go and mine, just to get rid of them, send them in the bush. No rules, no anything, and they created one of the worst environmental problems because there was no supervision. So, they just stopped them because of the environment. There was nothing they could do. They just stopped them. So, the Brazilians moved over to Suriname and French Guiana and they working there, but when the Surinamese government got serious about mining, which was only ten years ago, they clamped down on the Brazilians. So, the Brazilians, still pork knockers,[18] they came to Guyana. The French government now, they clamped down on mining. The people working in their country illegally mining had nothing to do with GGMC [unclear]. They got a special security working there, and when they catch you, you don't want to be there. So that is how we are now packed with Brazilians, and for the record, we will always be packed with Brazilians. There is nothing the government can do about it. Brazil is the only country that has our back. They are the only ones that we don't have a border dispute with. They are the only ones that recognize our borders, so they have our backs with Venezuela, as well as they have our backs with Suriname.[19]

Due to border disputes that remain as a legacy of colonialism, independent Guyana can do little to fight against the influx of small-scale gold miners from Brazil. Brazil is the only country with which Guyana does not have a border dispute, and it supports the country against other countries' attempts to claim land Guyana considers its own. The leading figure in gold and diamond mining in Guyana, interviewed in 2014, added that if "this idea of the government *ever* thinking they could stop mining, I say forget it. All you gonna do is create chaos. You gon' want a bigger army, because they can't stop the Brazilians from coming in, and if you got anything, you can't stop them. Brazilians is coming."[20] Gold mining, however, is the biggest driver of deforestation in these two countries.

Perhaps due to its importance as an economic activity and source of income in both countries, gold mining is a highly contentious issue around which different actors have widely conflicting interpretations and interests. Guyana and Suriname are no strangers to mining, since bauxite mining has taken place within their borders since the early 1800s, with some communities that developed around bauxite mining camps continuing to exist today. While central to both countries' economies, gold mining presents a major challenge to forest conservation efforts because of the increasing use

Figure 5. Small-scale miners in Mahdia, Guyana (Collins, 2014).

of more effective machinery by gold miners and the relative ease of movement in the starting up of operations. Gold mining is managed in Guyana by the Guyana Geology and Mines Commission (GGMC), and according to the laws on mining in Guyana, their permission is required to commence mining in Guyana. However, this procedure is not always followed. Mobile dredges are easy to set up, and the influx of Brazilian miners into the interior does not always follow the legal channels.

In recent years, gold mining has become the prime source of income, bringing in hundreds of millions of United States dollars in revenue yearly to the Guyana government. Newly developed institutional mechanisms in Suriname that seek to streamline what was previously a widespread unregulated small-scale gold mining sector there are also increasing state revenue. In Suriname, gold mining is the stronghold of Brazilian migrants and maroons, the descendants of enslaved Africans who escaped the slave plantations and set up communities in the forests. Even though small-scale gold mining reportedly yields a total revenue far greater than large-scale gold mining, the small-scale revenues had seldom come to the government, due to poor regulation. However, since 2010, the Gold Sector Planning Commission (OGS in Dutch) has attempted to create order in this area by working with some 20,000–30,000 illegal gold miners to streamline them and bring them into the country's formal systems. Due to the wide scale and long period of uncontrolled gold mining in the country, significant amounts of forests have been removed or degraded through these activities.

Nevertheless, both countries are planning to build or pave roads through the forests to connect to Brazil, which currently does not have direct overland access to the Atlantic Ocean and Caribbean Sea. These roads will cut through the depth of the forests, opening up the economic prospects for communities and increasing access, both local and foreign, to their resources. These activities do not necessarily jeopardize the payment for performance goals of REDD+, since payments were tied to remaining below a certain baseline of deforestation. In other words, these deforesting activities can continue and even rise within the REDD+ framework without putting possible REDD+ income at risk as long as they do not exceed the baseline of deforestation agreed on by the governments of Guyana and Suriname with REDD+ funders. The process of setting this baseline, however, is highly contentious.[21]

THE CLIMATE CHANGE AND DEVELOPMENT NEXUS

With climate change in mind and local demands for better standards of living in view, the state governments of Guyana and Suriname are therefore grappling with what I refer to as the climate change and development nexus. "The climate change and development nexus" refers to the set of circumstances where adaptation to climate change is deemed successful through improvements in infrastructure, which, in turn, depend on increased national income. However, economic activities with the potential to significantly increase the earnings of natural resource–rich countries tend to be extractive and environmentally harmful. These activities worsen the problem of climate change to different degrees, as seen with the major offshore oil discoveries in both countries after their involvement in REDD+.

In other words, the pursuit of infrastructural and economic growth development outcomes in Guyana and Suriname is highly likely to increase these states' contributions to greenhouse gas emissions and consequently to global climate change. As a result, climate change impacts on these vulnerable states will intensify, since economic development is likely to involve deforestation and the exploitation of other climate-significant natural resources, such as oil.[22] Both countries have increasingly demonstrated an awareness of the challenges posed by climate change in their policies, bringing development concerns to the fore. Guyana and Suriname's lack of preparation, in terms of infrastructure and resources for disaster response, however, goes on to fuel questions about the level of development necessary for averting climate change–related disasters.

FINDING A SOLUTION THROUGH REDD+

In May 2006, Suriname was hit by torrential rains that severely affected the forested areas of the country. Several major rivers of Suriname rose quickly and submerged surrounding areas. Some 22,000 people were displaced in the flood, and two-thirds of subsistence livestock and household goods in the area were destroyed.[23] Approximately 60 percent of the population along the Tapanahoni River in Suriname was displaced.[24] While heavy rainfall had become a regular consideration of life in Suriname, the rainfall of May 2006 was uncommon, necessitating the collective efforts of the Surinamese government and donor agencies.[25] Hence, questions about development that have persistently defined Guyana and Suriname's policy focus since independence were again brought to the forefront of national attention.

Similarly, a flood struck Guyana in 2005, exposing the coast's high vulnerability to flooding. Torrential rain caused the inundation of the coast on which the majority of the population resides. Complications in the country's water management system did little to ameliorate the situation, and the flood water remained stagnant for almost two weeks. Eventually, some relief came in the form of government interventions, donor assistance, and cooperation from the private sector. After a few days of chaos in which most economic life ground to a halt, the water receded. I witnessed firsthand some of the disastrous effects of the flood as people who were able moved their belongings to higher levels, storing them on tables or on the tops of refrigerators, waiting for the water to recede. Agriculture was devastated by the death of animals and crops, an additional vulnerability since the coastland possesses the most fertile land and is, hence, the center of the agricultural industry in Guyana. Schools were shut and offices closed as the water levels rose some 170 centimeters. Thirty-three people are known to have died from conditions related to the flood, with forty-nine hospitalized.[26] The provision of clean water was a priority, and health concerns grew due to an increase in the incidence of leptospirosis, which made a number of affected persons ill and claimed the lives of some others.[27]

Shortly thereafter, Guyana's then president, Bharrat Jagdeo, who led the country through the disaster response, made REDD+ central to his plan to have the country compensated for conserving the Amazon forests covering most of its noncoastal territory. He pointed to the vulnerability of Guyana's coast to sea-level rise, flooding, and other effects of climate change. In order to gain funding for these "development challenges,"[28] President Jagdeo, a trained economist, made what was then a radical proposal. He offered the services of Guyana's forests to the global community in an effort to remedy a "market failure" that allowed the world to benefit from the climate services of Guyana's forests without paying for them.[29] The proposed solution was that Guyana would be compensated by international actors for the work of its forests, presenting the option of simultaneous development and environmental protection. President Jagdeo positioned his argument in global terms, expressing himself in a measured and calculated way. This form of calculated expression, which has come to be central in the operationalization of REDD+ in both Guyana and Suriname, is evident in his statement that "solutions to deforestation are possible. They can be delivered quickly and cost effectively, and have the potential to transform the economic prospects of some of the poorest countries in the world."[30] The president set out to achieve interest and buy-in from actors in the international arena by highlighting the activities that take place in the country's forests that lead to deforestation, while drawing on the narrative of forests being worth more alive than dead, the embodiment of a market failure that must be corrected. In the policy paper outlining his position, he explained his aim:

> Assist those working within the UNFCCC [United Nations Framework Convention on Climate Change] process to deliver these solutions. It is built on the premise that much deforestation happens because individuals, communities and countries pursue legitimate economic activities—such as sell-

> ing timber or earning money and creating jobs in agriculture. The world economy values these activities. It does not value most of the services that forests provide when trees are kept alive, including the avoidance of greenhouse gas emissions. Correcting this market failure is the only long-term solution to deforestation.[31]

While portraying the market as an autonomous actor valuing and subsuming all aspects of the global environment, President Jagdeo sought to affirm that the forests in Guyana were not being held hostage. However, he pointed out that Guyana was in fact contributing little to deforestation. He suggested though that Guyana should be able to access funding to facilitate its economic development. Actual deforestation rates in Guyana were insufficient to warrant payments to remedy this perceived market failure. Therefore, the threat of future deforestation based on the needs of the development- and climate change–related needs of the country became the basis of problematizing deforestation. Through these efforts, REDD+ became the centerpiece for a new model of development in Guyana, one that intended to bring about infrastructural, economic, and other forms of development while addressing climate change through forest conservation. In neighboring Suriname, the conversation had not yet begun.

REDD+ ON THE INTERNATIONAL STAGE

Deforestation and forest degradation represent a significant source of global greenhouse gas emissions. They were estimated to represent 17 percent of global greenhouse gas emissions by the IPCC in 2007.[32] In 1997, as part of the global effort to combat climate change, the Kyoto Protocol had been established as a commitment taken by Annex I countries (thirty-seven industrialized countries and the European Community) to reduce greenhouse gases to 5 percent below 1990 emissions levels over the period spanning 2008 to 2012. However, there were no obligations for Non-Annex I countries, largely those countries classed as developing,[33] to change their development strategies. To provide flexible mechanisms through which Annex I countries could reduce their responsibility for greenhouse gas emissions, the Clean Development

Mechanism (CDM) was created as part of the Kyoto Protocol in 1997. This allowed Annex I countries to reduce their emissions outside of their borders through voluntary projects in developing countries. The carbon market grew out of this idea. While the CDM achieved deals for afforestation and reforestation, in the view of the governments of some highly forested countries, it was limited by the fact that it did not compensate for the conservation of standing forests.[34]

Forests, which were always seen as controversial within climate negotiations, had been left out of the climate change debate for some time. Policies aimed at conserving forests were seen in the international community as likely to encroach on the development aspirations of forested countries.[35] The difficulties associated with including carbon offsets from land use, land use change, and forestry (previously referred to as LULUCF) included limitations in measuring, reporting, and verifying the reductions. Some of these concerns have since been allayed by improvements in technology and methods of assessment in the subsequent development and implementation of REDD+,[36] such as improved remote sensing technologies.

Reducing emissions from deforestation in developing countries was first introduced at the Conference of the Parties to the UNFCCC in 2005 by a coalition of rainforest nations, including Costa Rica and Papua New Guinea. The group expressed a specific interest in considering the reduction of carbon emissions from deforestation and forest degradation in natural forests as a means of mitigating climate change. They formulated the original REDD (Reducing Emissions from Deforestation and forest Degradation) program to fill the gap of the CDM by providing payments for the work of standing forests. When the program was expanded to compensate both efforts, to prevent emissions and to increase the removal of carbon from the atmosphere,[37] it became known as REDD+.[38] The 2010 Cancun Agreement reflected this progress on the thinking behind REDD by broadening the scope of REDD and addressing the criticism that the original concept offered safeguards only for those forests that were at immediate risk of destruction while ignoring those that had been studiously conserved and protected for years. The Cancun Agreement also featured environmental and social safeguards geared towards forest conservation and respect for the rights of indigenous communities and traditional knowledge, so countries participating in REDD+ were required to ensure that social and environmental safeguards would be addressed in their national strategies.[39]

The REDD+ mechanism was not lauded by all. Numerous concerns about the initiative have been voiced regarding issues such as its effect on the indigenous populations who reside within the forests,[40] land tenure and legal issues,[41] the ability of REDD+ countries to transparently manage large influxes of cash, and measurement practices.[42] Nevertheless, an international consensus on the implementation of REDD+, which was eventually included in the Paris Agreement, was adopted in 2015.

REDD+ has been subject to intense continued debate centered on its ability to generate substantial, tangible results.[43] Angelsen, one of the earliest supporters of incentivizing conservation through markets, recently argued that REDD+ is more similar to results-based aid given at will by wealthier countries than it is to a market-based instrument paid on performance. Using Guyana as one of his case studies, Angelsen argued that REDD+ itself has been reinterpreted within international finance practices, in national policies, and on the ground in ways that are synonymous with largely ineffective conservation efforts.[44] He asserted that this reinterpretation, coupled with the fact that both REDD+ and the carbon market that is intended to support it have still not been fully realized, contributes to REDD+'s lack of results in stemming the tide of deforestation globally.

However, although REDD+ was eventually adopted into the United Nations mechanisms for addressing climate change, it remains a loose amalgam of efforts aimed at identifying funding sources, testing its principles in different localities around the globe, and preparing candidate countries for its implementation. Overall, REDD+ seeks to create incentives for avoiding deforestation. It represents a contribution that forested, developing countries could make to mitigating climate change while continuing to develop in low-carbon or environmentally benign ways.

REDD+ IN GUYANA AND SURINAME

REDD+ is characterized by a strong emphasis on demarcating aspects of the natural environment based on measurable functions such as carbon storage and land value. It relies on technologies through which the forests of participating countries are monitored remotely, as well as on economic valuation of aspects of the environment. National government representatives tend to emphasize the potential of REDD+ for bringing development to their countries. Further, their participation in REDD+ is justified, in the cases of Guyana and Suriname, through narratives centered on the need for climate change readiness, the fulfillment of basic needs for education and health care, the strengthening of the system of allocating land rights, and the need for economic growth, albeit in the different guises of green or low-carbon growth.

In 2009, Guyana signed the Guyana-Norway agreement committing Norway to providing up to US$250 million to Guyana as payment for avoiding deforestation. This amount represented a woeful 11.6 percent of the amount the government had originally identified as the opportunity cost of not exploiting Guyana's forests. Guyana, through its bilateral REDD+ agreement with Norway, was intended to stand as a global showcase of how climate change can be addressed through low-carbon development and international cooperation. Guyana also received US$3.8 million from the Forest Carbon Partnership Facility (FCPF) to support its REDD+ readiness process, along with smaller grants from nongovernmental and intergovernmental organizations. Guyana's REDD+ preparation progressed slowly, with its readiness preparation proposal (R-PP), required to source funding for preparation activities, being approved in 2009. Thus far, only forest conserved and managed by the state (amounting to approximately 80% of the nation's forest cover) has been allocated for REDD+ in Guyana. Plans should then have been developed for indigenous communities with legally defined land rights, to give them the option of including their titled forests.

Suriname was not one of REDD+'s early movers. However, it eventually started to pursue a national REDD+ initiative, having joined the global effort led by the World Bank and the United Nations REDD+ Programme (UN-REDD+) in 2009. While its entrance to the realm of "avoided deforestation" through incentives was not as dramatic as that of Guyana's,[45] its process of preparing for the implementation of REDD+ also unearthed a number of challenges that highlight the difficulties of instituting this market-based conservation logic. Central amongst the issues faced by

Suriname was the resistance of the indigenous and maroon communities living in the forests. After reading about the initiative online, they felt that they had not been adequately consulted about this use of forests they considered their own, and they challenged the government's lack of consultation with them in planned national REDD+ activities.[46] Suriname was eventually able to get its R-PP approved by the World Bank in March 2013, after it had been rejected twice on account of lobbying by communities living in the forest. Suriname also received US$3.8 million from the World Bank to support its REDD+ readiness efforts and was subsequently awarded an additional US$2.6 million to complete its preparation process.

REDD+ in Suriname was presented as a means of planning sustainably, as part of a climate compatible development strategy (CCDS) that the Suriname government had begun to conceptualize. The CCDS had aimed to minimize the impacts of climate change while maximizing opportunities for human development towards a more resilient future. REDD+ was intended to seek out a development path that balances social, economic, and environmental issues and to function as a tool for sustainably managing the forests without limiting Suriname's economic and social development.[47] Across both countries, nongovernmental organizations such as Conservation International and the World Wildlife Fund, civil society organizations representing marginalized groups, and government offices with varying foci worked in a coordinated way towards REDD+ readiness with collaborative efforts taking place across the Guiana Shield.

THE COLONIALITY OF REDD+ AND NEOLIBERAL CONSERVATION

REDD+ is an iconic representation of a certain set of assumptions of the relationship between markets and nature, depicting the conservation of forests as a charitable ideal, which should be superseded by rational, calculated decisions, as Jagdeo's rhetoric demonstrated. REDD+ also assumes an actor with the power to choose cutting down trees as one option among many, like the variety of food options at an all-you-can-eat buffet. The aim of this process is to incentivize this actor to opt for the non-deforesting option, which should ideally be long-lasting, satisfying, and furthering of the global good.

REDD+ is grounded in a strongly economic and rationalist approach to decision-making. Its focus on the creation of incentives in the aim of reducing deforestation assumes that it is possible to provide incentives against present or future deforestation. It also makes the actors it targets responsible for the chain of activities that result in deforestation. When the incentive is financial, as in the case of REDD+, it ignores the origin of the funding for the incentive as long as the conservation of a particular demarcated area of forests is achieved. It subsumes environmental and developmental aims under the umbrella of the managerial structure provided by the market and assumes that these aims can be reached simultaneously, while sidestepping the inherent trade-off between the two goals. In a nutshell, it is exemplary of the increasingly commonplace logics of neoliberal environmental conservation.

In general, environmental conservation is portrayed as a counter of sorts to capitalism's environmental excesses.[48,49] Since its relatively recent emergence, the global conservation movement has constantly shifted, moving from a top-down, punitive fortress conservation model to a more grounded, community-based approach.[50] In the past few decades, however, conservation came to explicitly embrace the principles of the market as a guiding rationale by taking on forms such as "ecotourism, payments for ecosystem services and biodiversity derivatives" along with other aspects of financial and technological instruments.[51] Conservation's increasing neoliberalization reflects the deepening and spread of neoliberalism that has since resulted in a reduction of public funding for conservation globally,[52] to which conservation organizations responded by turning to the market.

Capitalism, currently the dominant economic system, had its origins in the sixteenth and seventeenth centuries. However, neoliberalism, an instrument of capitalism's intensification, commenced in the 1970s, modelling social relations and public affairs on the discipline of capitalism.[53] Capitalism draws on and necessitates the principles of commodification, market discipline, competition, and financialization.[54,55] Far from being internally coherent,[56,57] neoliberalism is characterized by market centrality, a shift in the role of the nation-state and its functioning in the post–World War II era, and Anglo-American roots.[58] It holds superior individualized, market-based competition.[59] In poststructuralist thought, neoliberalism represents a new form of governmentality that traces how "hegemonic" ideas of market centrality become inscribed in everyday practices.[60,61] In essence, neoliberalism is characterized by its assertion that human well-being is best served by maximizing individual and entrepreneurial freedoms through a system that prioritizes property rights, unfettered market access, and trade. The role of the state in this system is to set up the conditions required for the operation of the market but to stay out of the market's operation and effects.[62]

Payments for ecosystem services (PES), as a suite of policy instruments of which REDD+ is emblematic, are market-based instruments for managing the environment. PES is a form of neoliberal environmental governance through which "state-centered 'command-and-control' policies are intended to be replaced by 'market-based instruments' (MBIs) seeking to incentivize sustainable resource management in the absence of direct regulation."[63] The marketization of conservation has been deeply studied within the subfield of political ecology called neoliberal conservation, which examines the marriage between conventionally damaging market activities and the goal of conserving the environment. PES is critiqued in this scholarship on the grounds that its solutions for environmental concerns rely on the same activities that prompted those concerns in the first place.[64] In this vein, neoliberalized nature is seen as expanding the reach of capital by financing and commodifying the conservation of previously public nature.[65,66]

REDD+ represents the largest and most ambitious PES effort towards fixing what is seen as a market failure, that of nature remaining financially undervalued in capital markets. REDD+, in this sense, is characterized by the idea that environmental problems emerge because prices inaccurately reflect the cost of producing and using fossil fuels and overusing natural resources. It depicts these considerations as market externalities. According to this logic, the appropriate response is to fix and regulate the market by putting prices on environmental services and integrating them into the market apparatus.

However, while the neoliberal conservation literature has consistently called attention to the varied challenges and weaknesses of market-based instruments, it has not paid significant attention to how racialized dynamics rooted in colonial histories affect their implementation.[67,68] Many of the areas prioritized as needing to be conserved are located within formerly colonized spaces, with most conservation funding coming from (former) colonial centers. The coloniality embedded in this system can thus be seen through the structure of these financial flows and their accompanying power asymmetries between Northern and Southern countries. These asymmetries are manifested in ways that "shape interactions between dominant and marginalized stakeholders across organizations, institutions and scales."[69] These power relations "intersect and privilege certain ways of understanding the world that often perpetuate entrenched colonial pretension to truth, that is, which knowledge claims are deemed truthful and by whom the truth can be declared."[70] They can also be seen in how Eurocentric thought, priorities, and approaches are often privileged even by those who are not European, and in the marginalization of indigenous knowledge systems. In these ways, neoliberal conservation initiatives, like REDD+, risk "exacerbating these on-the-ground inequalities and inviting conservation failures."[71] Therefore, colonial histories and their racialized grammar can indeed be seen, in the words of philosopher James Trafford, to be "fundamental to the organizing structure of our worlds, with its neoliberal organization just one iteration of ongoing coloniality."[72]

CONCLUSION

This chapter has conceptualized the pursuit of REDD+ within what I referred to as the development and climate change nexus. It drew attention to how developing countries facing the debilitating impacts of climate change interpret and repackage this challenge through their socioeconomic realities. Deforestation, REDD+'s primary focus, is situated within these competing and overlapping demands, that is, between the need to gain funding to facilitate development and the need for these countries to prepare themselves for climate change. However, the environmentally exploitative development trajectories through which countries like Guyana and Suriname primarily earn an income, to then be directed towards adapting to climate change or improving development outcomes for their constituents, tend to worsen the problem of climate change and ultimately their own fates.

Some critiques of REDD+ justifiably challenge the colonial and ethical implications of largely wealthy Western countries paying poorer Southern ones, often their former colonies,[73] to conserve their forests. However, as I argued in this chapter, this critique, while well founded, is too blunt. I ask instead, "When is it valid to question the motives and actions of postcolonial, climate vulnerable countries that worsen their own situations in relation to climate change through deforesting and other activities meant to benefit their populations? How do we question their development strategies without automatically recentering industrialized countries in the debate on attributions of responsibility?" These questions come with the risk that I am shifting blame from the feet of industrialized countries whose actions brought about the problem of climate change and continue to contribute significantly to it. However, my focus here lies in highlighting that the development trajectories undertaken by not-yet-industrialized countries like Guyana and Suriname need to be altered to meet the demands of the climate-changing era in which we live.

As pointed out by Juno Salazar Parreñas, decolonization demands "a rejection of a telos."[74] In this case, decolonization requires the rejection of a Western-centered modernization telos and linear narratives of environmentally harmful "development." Hence, decolonizing environmental governance in Guyana and Suriname must call into question the presumption that some imagined stage of development, whether achieved through gold mining, oil production, or other deforesting activities, is the desirable end goal. It requires that efforts to imagine new paths be taken seriously. This chapter and those that follow it continue in this vein. They collectively argue that as climate change becomes increasingly felt and perceived, development trajectories must be meaningfully rethought and reexamined, or altogether eschewed. Such is the decolonial horizon.

2 Behead the Sovereign

> We have allowed the shadows of our history to dominate our potential and to perpetuate fear. We have become captive to our own racial and political stereotypes.[1]

> Natural riches and resources are property of the nation and shall be used to promote economic, social and cultural development. The nation shall have the inalienable right to take complete possession of the natural resources in order to apply them to the needs of the economic, social and cultural development of Suriname.[2]

The independent states of Guyana and Suriname are almost entirely colonial constructions. Forest-dependent people, especially those indigenous to the region, have historically resisted the process of state formation that emerged from the violent disruption of their pre-colonial, forest use traditions. Through these disruptions, lands through which they had once freely roamed became European colonies; societies and environments shaped by colonialism were racially stratified; the sovereign state governments that took over after independence were imbued with the power to issue concessions for extractive activity in lands communities considered their own; and independent governments were empowered to allocate the carbon stored in state forests to REDD+ and global carbon offsetting ventures.

Through European colonization, the Caribbean region, of which Guyana and Suriname are culturally and politically a part, came to function as a backdrop of various meetings of culture and ethnicity. Over time, there emerged "one of the most diverse yet intricately interconnected geopolitical and cultural regions in the modern world."[3] The fruits of this interconnectedness were deeply disadvantageous to particular groups of people and the natural environment. Few places were more detrimentally impacted by European expansion than the Caribbean.[4] The indigenous Caribbean population, for instance, was devastated at a scale unseen elsewhere. The environment was negatively affected as small-scale farming was displaced by large-scale plantation agriculture. As a corollary, a constant demand for labor emerged.[5]

This chapter traces this formative history of Guyana and Suriname while highlighting the twin processes through which, on one hand, racial and political stereotypes became entrenched in these societies and, on the other, natural resources became the inalienable property of the state. Some of the racial and political stereotypes shaped in response to colonial rule include the agricultural East Indian indentured servant; the vulnerable forest-dwelling Amerindian; the enslaved, liberated, and gold-mining African; the trading Portuguese; and the rich, white European master. These subject positions developed in tandem with the sovereign, colonial state as outcomes of racialized relations to capital-mediated environments. These positions are rooted in five centuries of colonialism that continue to structure the societies of Guyana and Suriname today.

Meanwhile, the colonization of the lands that eventually became the territory of Guyana and Suriname was carried out on account of the ability of the land to generate economic surplus to be repatriated to the colonial metropole. This economic surplus intensified processes of capital accumulation taking place in the colonial metropole.[6] In the colonies, however, this colonization fostered and inculcated an ethic through which local populations came to see their lands, lives, and environment as stores of value to be exploited for external consumption.

Consequently, attempts to decolonize environmental governance in states constructed through centuries of colonization depend on recognizing and challenging the automatic nature through which the sovereign head of the state is deemed responsible for governance. In this chapter, I point to "the head of the sovereign" as a representation of statehood rooted in European colonization and argue that figuratively beheading the sovereign would allow for nonstate actors to be increasingly recognized for their ability to effectively manage and govern forests. Consciously reducing deference to the sovereign state in the awareness of its colonial grounding forms a necessary step towards making space for other forms, logics, and mentalities of being and governing to take root.

BECOMING SOVEREIGN—GUYANA

Before the Sovereign

Indigenous Amerindians inhabiting the lands that became Guyana's territory were confronted in the early 1600s by Europeans, who violently disrupted their forest use traditions. This period of disruption marked the origin point of sovereign colonial governance in Guyana and its forests.

Inspired by Foucault,[7] and as highlighted in the introduction of this book, I use the term *(de)colonial mentality* in this chapter and throughout the book to signal the overlapping nature of governing strategies of colonization and decolonization—in other words, (de)colonization. I refer to the colonial mode of governing the forests that unfolded after this initial disruptive encounter as "sovereign" (de)colonial mentality. This sovereign mode of governing operates by imposing formal rules and regulations on the land, forest, and human subjects it seeks to govern. As a result, this governing strategy remains an important area for decolonization in the post-independence period.

While the history of the Guiana Shield prior to colonization is poorly documented,[8] the Amerindian communities residing in the area that became Guyana were known to be numerous and diverse. Amerindian peoples occupied those lands for several millennia before the arrival of the Europeans in the sixteenth century.[9] These indigenous groups were nomadic and moved freely across the Amazon basin.[10] The Warraus, for example, were said to be boat builders who foraged and fished for their meals, while the Arawaks were said to prefer higher ground along the rivers and to have practiced agriculture on crops tailored to their environments. Of the several Amerindian groups in Guyana, the fierce, warrior-like Caribs were the most dominant at the time of European arrival.[11] Their societies were "well-ordered and technologically complex hierarchical societies based on intensive agriculture and fishing."[12] Presently, there are four main groups in Guyana: the Warraus, Arawaks, Wapishanas, and the Caribs, which have formed smaller groupings of Arrecunas, Akawaios,

Patamonas, Macushis, and Wai-wais.[13] After European arrival, Amerindian societies began to take on noticeable changes as they adopted European technologies that were better suited for clearing forests and building houses. The Europeans, for their part, established trading relationships with the Amerindians that were based on the system of barter, exchanging industrial goods for food and forest products.

The Colonial Sovereign in Guyana

The arrival of the Europeans to the Amazonian Guiana Shield, within which Guyana and Suriname are situated, was followed by several battles and conquests directly tied to events in Europe. These battles saw the colonies that became independent Guyana—Essequibo, Demerara, and Berbice—change hands on numerous occasions, being controlled by the Dutch, the French, and the British during more than 400 years of colonial rule. The borders between the now-independent Guyana and Suriname countries had also changed occasionally as the colonies were amalgamated or divided into parcels of land to be governed by the colonial power du jour.

However, this colonization depended on a sustained supply of cheap, available labor. The Dutch established their first permanent settlement in Guyana in 1616 and called it Kykoveral and eventually set up sugar plantations on the coasts. The labor on the plantations was first provided by "red slaves," as Amerindians who had been captured by other Amerindians were then called. The plantations were also reliant on indentured European labor. Between the sixteenth and nineteenth centuries, a European indentureship system emerged, rather than a system of freely provided labor, because potential laborers often found themselves unable to finance their journey overseas despite their willingness to relocate. Hence, contracts were drawn up between prospective migrants and the shipping firms or employers. These contracts stipulated that the migrant laborer would indenture him- or herself for a defined period, often between three and seven years, during which the laborer would work for the employer in exchange for travel, housing, and subsistence for the period of the contract.[14] Here, the reader should recall that at the time, Europe itself was in a process of transition from feudalism to capitalism.[15]

It was not the relatively coherent, "modern" region and political entity it is thought to be now.

Many of the early indentured laborers brought to the Americas from Europe were coerced, with both voluntary and involuntary European labor coexisting during some periods. Involuntary servitude was often forced onto those deemed undesirable in Europe at the time, including vagrants or vagabonds, the poor and exploited underclass, felons, prostitutes, and even children and teenagers from Ireland and Britain.[16] This practice continued into later centuries as "roughly 10,000 Scottish, English, Irish, and even German prisoners from the 1651 Battle of Worcester, the final battle of the English Civil War, were also transported to the Americas as servants."[17] Those who were captured through war, or were deemed political prisoners, could be sold for up to ten years of service, a period significantly lengthier than the customary five-to-seven-year period.[18]

Over time, indentureship grew to become "the most dependable source of coerced labor across the British American colonies."[19] It remained a key source of labor for plantations in the Americas throughout the seventeenth and eighteenth centuries. Although estimates differ, more than half a million Europeans are estimated to have been employed through indentureship in plantations in the Caribbean. By the eighteenth century, small farms were increasingly being overtaken by large plantations, driving up the demand and cost of indentured European labor so dramatically that colonizer preferences shifted towards less expensive, seemingly more suitable, and abundant enslaved Africans.[20] This change in preferences led to millions of enslaved Africans being forcibly transported across the Atlantic,[21] with significant numbers arriving at Guyana's shores.

Through this shift in the European colonizers' preference in labor sources, a massive system of enslaved labor developed at a scale never before seen, as European colonial powers sourced and shipped enslaved Africans for work in plantations in the New World. The enslaved Africans effectively added a third major group of people to Guyana's then-burgeoning multiethnic status. While the enslavement and trafficking of enslaved persons was by no means a new phenomenon at that point in human history, the transatlantic slave trade was significant for becoming the "last and greatest of all slave trading systems."[22] Approximately 12.5 million Africans were sent to the New World, where they were made to labor in appalling conditions. Most of those enslaved and brought to the New World disembarked in European colonies in the Caribbean, including then Guiana. This system depended on the involvement of a wide cross section of actors and powers across the Americas, Africa, and Europe and a considerable amount of risk and navigational prowess in moving those enslaved, thousands of miles across the Atlantic and Indian Oceans.[23]

The Dutch colonizers engaged heavily in the transatlantic slave trade. As a result, the arrival of enslaved Africans to work on the plantations brought a new role for the Amerindians. The Amerindian trading allies of the Dutch became "owls," or guards of sorts, as the Dutch began to reward Amerindian allies for capturing enslaved Africans who ran away to seek refuge in the forests. By the 1760s, the norm was that the Amerindians would fill the role of policing the interior and of providing weaker Amerindians, known as red slaves, to work on the plantations for the Dutch. This was not the case in other Caribbean territories like Suriname and Jamaica, where successful maroon communities were established and maintained, in some cases, through collaboration with indigenous communities.

The number of enslaved Africans brought to Guyana kept increasing as European colonial powers sourced and shipped enslaved Africans for work in plantations in the New World. Meanwhile, the alliance between the Dutch and the Amerindians in Guyana remained strong. This was evidenced when, in 1763, enslaved Africans on the plantations in Guyana rose up to fight for their freedom. The Amerindians accepted an offer of arms from the Dutch and helped them to put down the rebellion. Their alliance only weakened as the Dutch demand for Amerindian red slaves waned, due to the increasing plantation labor supply from enslaved Africans. The Amerindian slave trade was abolished in 1793, and the trade relationship of the Amerindians with the Dutch finally came to an end.[24]

Nevertheless, these events paved the way for a future relationship of distrust between Amerindian communities and enslaved Africans and their descendants, who were becoming all the more populous on the coasts while the Amerindians themselves were shrinking in number due to war and disease in forests. In 1803, due to changing power relations between European states, the three Dutch colonies of Essequibo, Demerara, and Berbice, comprising modern-day Guyana, passed to British control.[25] The colonies were united in 1831 under the banner of British Guiana. The relationship between Amerindians and Guyana's new colonizer relegated Amerindians to the role of bush police.[26] Enslaved Africans were eventually emancipated in 1833 on account of a triad of shifting public opinion in England, coordinated uprisings in the colonies, and the possibility of having goods produced at a cheaper cost elsewhere. In the forests of the then-British colony of Guiana, however, emancipation had the unintended consequence of further obliterating the role of Amerindians.[27]

The emancipation of enslaved Africans in 1833 reduced the ready labor supply dramatically. In an attempt to remedy the situation, the British colonizers returned to indentureship. They brought large groups of people to Guyana from China in the 1860s, from Portugal in the 1880s, and most notably due to their large numbers, from India in the 1830s. Land issues emerged as a challenge in this context as the colonial authorities sought to make alternative forms of livelihood difficult for the freed slaves. In Guyana, small-scale or artisanal mining became one alternative livelihood source of income for African descendants. Pork knockers emerged as a group of Afro descendants who traveled to the interior locations of the forests with a bag of rice on their backs, a barrel of pork, equipment to hunt for "wild meat" to survive, and a shovel. Pork knockers would camp in the forested interior and mine at a small scale.[28] As the prices for machinery later became more within the reach of these miners, they started to use mechanized methods of extracting gold, which facilitated the digging of bigger holes and the felling of more trees. This historical chain of events would have significant consequences on later forest conservation efforts through REDD+.

Colonial authorities deployed a strategy of divide and rule to foment strife between the different ethnicities to prevent them from collectively organizing.[29] As Ross (1996) highlights:

> The coolie despises the negro because he considers him . . . not so highly civilized as himself, while the negro . . . despises the coolie because he is so immensely inferior to himself in physical strength. There will never be much danger of seditious disturbances among the East Indian immigrants . . . so long as large numbers of negroes continue to be employed with them.[30]

Divisions between individuals of East Indian descent, derogatorily referred to now as "coolies,"[31] and those of African descent, referred to as negroes, resulted from a conscious policy of managing the multiethnic labor in such a way that it maintained European dominance in the society that emerged in British Guiana. This society was framed by these labor relations. At the top of the social order were white planters, who had the support of the colonial authorities and controlled the majority of land and capital. Chinese and Portuguese, freed from the restrictive laws and able to acquire land, emerged as a class of vigorous smallholders practicing market gardening, which provided the basis of their later prominence in commerce, charcoaling, and gold. Free villages of poor former enslaved Africans eked out a living on marginal lands or worked as seasonal labor on the plantations, to which East Indian indentured and formerly indentured laborers were more closely tied year-round. Altogether, over ninety percent of the population was concentrated along the narrow coastal strip, while the Amerindians, out of sight and out of mind, continued to populate the forested interior.[32] It is important to note that there was no inherent tension between East Indians and Africans, but it was the organization of the colonial society that created and fostered this strife.

At the end of the nineteenth century, the immigrant labor began to reorganize themselves. A black middle class emerged in urban areas through education and public and administrative service. East Indians expanded their economic base in the rural areas through rice farming on land granted to them by the colonial masters in exchange for their choice not to demand their return passage to India. The outcome was an emergent black middle class and an East Indian dominance over rural rice-producing areas as groups sought to find ways to sustain themselves in their new homelands.[33] The divide-and-rule policy was not limited in effect to coastal populations, however. It also colored the interactions between indigenous groups and the remainder of the society in Guyana, as a legacy of the former's role as gatekeepers to the forest, preventing runaway slaves from seeking refuge there. This, and the subsequent separation of Amerindian groups from the society spatially (due to their primary residence in the forested areas and outside of the more "modern" coastal areas),[34] led to these forested communities being represented as backward, unmodern, and in need of help from external sources.[35]

All in all, colonial rule in Guyana was a clear expression of sovereign (de)colonial mentality through which sovereign rulers in Europe and their agents in the Americas violently exercised a practice of taking life or letting live by reordering and making claims on human bodies, land, and forests.[36] While other (de)colonial mentalities were likely present during this colonial period, they were secondary to the violence and force of the colonial sovereign to manage the population and the environment.[37]

Later, another major disruption to the strategies of governing Guyana's territory and forests took place when in 1966, the colony gained independence from the British to form an independent state.[38] Locals, by then violently and coercively brought to this area from other continents to join indigenous inhabitants, were left to govern themselves and the forests. The colonizer, in fleshy form with violent, imperial backing, went home. However, the focus of successive governments of independent Guyana, including previous colonial authorities, remained on the exploitation of timber and mineral resources through the racialized immigrant labor that remained to form the Guyanese population.[39]

The Independent Sovereign of Guyana

The relocation of large groups of people from around the world to colonies in the Americas was spurred primarily by European influence and the need for labor on the plantations. As a result, the economic value the colonizer placed on Guyana's land, people, and natural resources became a determining factor in its past and future. Guyana's political, social, and economic situation is still reeling from its past of slavery and servitude. This is demonstrated by ethnic polarization and strife evident in the country where political parties in Guyana continue to draw their support from ethnic groups.[40] The population continues to identify themselves along racial fault lines, demonstrated by voting patterns, areas of residence, and sources of income. After independence, sugar continued to play an important but dwindling role in the Guyanese economy despite attempts by the independent government, foreign companies, and small local producers to diversify the economy away from it. The majority of the population continues to reside near the coastal sugarcane plantations. The cities are populated by the descendants of enslaved Africans while the rural population is dominated by the descendants of East Indian indentured workers.[41]

The People's Progressive Party (PPP) emerged in the pre-independence period while championing an ethic founded on opposition to the exploitative practices of the transnational sugar corporations, and independent workers' rights and socialism. In order to hand independence to a more malleable governing body, Britain and the United States fostered racial and political violence lasting from the mid-1950s to the mid-1960s. The PPP leader, Cheddi Jagan, was vilified publicly and jailed for six months by the colonial government, who opposed his rhetoric. His imprisonment resulted in reduced confidence from supporters. The party then split, and the People's National Congress (PNC) was created. Britain and the United States then supported the opposition PNC as part of the Cold War maneuvers to ensure that another socialist state was not established in the Western Hemisphere.[42] This division later took on racial tones with Indo-Guyanese overwhelmingly supporting the PPP and Afro-Guyanese supporting the PNC.[43]

In 1988, after decades of PNC corruption and mismanagement of the economy, along with fraudulent elections, it became clear that socialism had failed in Guyana. The new PNC leader of the country, Desmond Hoyte, embarked on structural adjustment and development based on foreign investment. He liberalized the economy, pushed to intensify the exploitation of natural resources to stimulate development, eliminated price controls, permitted a floating exchange rate, reduced import tariffs, and privatized state assets. These were just some of the neoliberal policies he adopted to reduce state intervention in the economy and to set up a self-regulating market as part of the world economy. With the support of the World Bank, Guyana's debt repayments were eliminated, and debts were forgiven. This significantly freed up budgetary allocations for development.

In 1992, the PPP again took the reins of government and continued to institute widespread market reforms. Influenced by the International Monetary Fund (IMF) and World Bank, the PPP embraced foreign investment as part of structural adjustment programs. Jagan, however, encouraged South-South relationships and tried to break Guyana's dependency on Western development sources.[44] To date, however, both major political parties draw their support from ethnically affiliated supporters. Voting in Guyana continues to be carried out along ethnic lines, with the PPP being predominantly supported by rural Indo-Guyanese voters, and the PNC, now part of a coalition with smaller ethnically mixed parties, supported by largely urban Afro-Guyanese voters.

Racialized Identities and Environments

> The majority Afro- and Indo-populations have been administered as types of "racial" citizens.[45]

In Guyana, the representation of different ethnicities came to be imbued with a particular role in the economy and in relation to the natural environment. Within the colony of British Guiana, East Indians had worked largely as paid laborers on the sugar plantations, while establishing their own agricultural ventures in rice. African descendants were involved in mining bauxite in the areas of Mackenzie and Kwakwani, some miles outside of Georgetown. African descendants and mixed-race individuals dominated the civil and educational sectors. The Chinese and Portuguese groups within British Guiana derived their sustenance from trade.[46] This economic separation is depicted by the following quote extracted from an interview with a leading figure in gold and diamond mining in the now-independent Guyana:

> Mining was distinctly divided in terms of who was doing what. The Afro-blacks, they were the pork knockers. They did the digging; they did the mining and they brought out the gold. The Portuguese were the pawnbrokers, pawn the gold and sell the jewelry. The East Indians were the goldsmiths. They made the gold. They used to buy the gold. As the price of gold went up, the first people who got the bright idea was the goldsmiths. Why would they buy gold? So, they started to invest in mining, so they started to go into mining as a question of business. The Portuguese were always backing. They weren't the pork knockers. They would back the pork knockers. You have to understand.[47]

The experience of the Portuguese is notable, since they were brought to British Guiana as immigrant labor in an attempt to increase the colony's white population and to stave off some of the slave rebellions that were taking place in the years before emancipation. The Portuguese were never seen as equal in status to the British colonial masters, and they themselves resented having to work on the plantation alongside the now-free Africans, who they saw as inferior beings.[48,49] This racialized separation of labor lives on in Guyana today, where people speak of the land of six peoples: Amerindians, Europeans, Africans, East Indians, Chinese, and the Portuguese. Note here the distinction between Portuguese and Europeans, since the two were seen to be different "races." This difference was based not primarily on the color of their skin, but on their economic status and societal position of power on arrival to the colonies.

An interview I conducted with a dominant figure in regulating the mining industry confirmed these racialized roles in relating to the environment mediated by the economy that emerged after independence. While acknowledging that all pork knockers were indeed of African descent, he stated that "when the laws were done, the British made sure that when they had the indentured servants come in, that the indentured servants couldn't leave to go elsewhere without permission, so there was a statement in the mining laws that the East Indians had to get permission to go into the mining areas."[50] He further described that the Portuguese and Chinese were unable to venture into the forests either and that these racial divides play a role in how people manage their resources today. He continued:

> What I know for sure is that a lot of the [mining] claims, which are the small-scale holdings, were held by people of African descent for many, many years. Gradually, others bought from them and then others located their own claims. In 1994, we had the introduction of medium scale, and that medium scale had a key element that it was going to make available to the Guyanese, land without them having to go into the bush [forests] to get it. If you wanted a claim, you had to go into the bush, prospect, and then claim the area that you had prospected. That is why it is called a claim. To get the medium scale, you could go on a map, identify it on a map, and apply for it by geographic description only, so we had a lot of people doing that, and the monied class then stepped into the industry in a much bigger way.[51]

These racialized identities were taken a step further by one policymaker working on providing payments for ecosystem services (PES) activities in both Guyana and Suriname. He saw the internal divisions of Guyana and Suriname as the root of the incoherent approaches of these countries in managing their natural environment. He stated:

> In Guyana and Suriname, we have a number of cultures living together that brought different ideas from different parts of the world, and we haven't yet melded a nation from a coming together of these cultures and traditions. Finding out what I would say the best aspects of these cultures and traditions, that are essentially compatible with nature, I haven't seen any of the cultures that have been brought here, or any of the traditions that have been brought here, that are essentially nature destroying in nature. They are not nature destroying at all. In fact, what has been nature destroying has been this tendency to respond to markets, and these are external markets which requires "Gold? The price is high. OK, we have it. We take it out. We sell it. . . ." We disrupt the environment. We disrupt the equilibrium. We need to establish a new equilibrium through some managed activities. We seldom do that. We depend on nature to deploy its healing properties.[52]

Pointing to the diverse people and cultures comprising Guyana and Suriname, this interviewee argued that deforesting and environmentally harmful activities are not due to the cultures themselves, but to the response of the people in the countries to the demands of the externally driven market and due to the lack of a melded national identity. The satisfaction of these demands results in local environmental destruction, which is seldom actively remedied by people in either country. In Guyana and Suriname, the demands of the international arena are met with little concern for its local effects.

In the 1950s, the East Indians formed the largest group of indentured servants in the colony of British Guiana, numbering approximately 250,000. They kept working on the sugar estates, and some of their descendants eventually took up rice farming.[53] The smaller numbers of indentured servants made up by the Portuguese and Chinese stopped working on the plantations as soon as they were able to and formed a stratum of peddlers, shopkeepers, and urban tradesmen and professionals.[54] Lowenthal commented on race relations in Guyana by sharing a discussion to which he was privy. He reported a conversation that took place in the 1950s: "'The trouble with us,' a Portuguese explained to me in Georgetown, 'is that we are not a nation. If you ask a man here what he is, he will tell you. I am Creole, I am Indian, I am Portuguese, I am Chinese; never Guianese.'"[55] Hence, although different cultures brought to Guyana and Suriname adapted to their new environments and circumstances, they have adhered largely to racialized roles available to them in the colonial period based on the circumstances of their arrival. These roles were supportive of capital accumulation and continue to be instrumental in both countries today.[56]

The small-scale and artisanal gold mining industry of the post-independence period was also almost entirely characterized by racialized identities. By this, I mean that one particular ethnic group was mainly responsible for different aspects of the process. By the 1900s, there were approximately six thousand black pork knockers in Guyana's interior who adopted simple techniques that allowed them to earn and establish small communities in the forests. Being deprived of access to land and employment on the coasts, African descendants were willing to risk their fortunes in the interior of Guyana in search of gold.[57]

Since then, small-scale gold mining in Guyana has experienced many periods of peaks and troughs. In the 1970s, it experienced a boom spurred by a rise in gold prices and the introduction of mechanized mining that allowed for much easier and more widespread mining, which was, in turn, more ecologically destructive. Then, in the 1970s and 80s, Brazilian miners started to flow across the border with improved technologies that further fed this destruction.[58] Gold mining, in effect, became the mainstay of the descendants of enslaved Africans, with profound negative environmental and social effects on the indigenous communities already residing there.

After it was discovered in the late 1870s, bauxite also formed an important part of Guyana's effort to diversify economically. The Aluminum Company of America (Alcoa) became the aluminum monopoly in Guyana, supplying the United States with high-quality bauxite, which was known as "red gold" during the Second World War and the postwar years. Guyana's supply of bauxite to the United States was complemented by bauxite sourced from neighboring Suriname. These economic and natural resource extractive activities, however, generated little financial gain for British Guiana, since Alcoa had bought the land from its previous owners at an exceptionally low cost under the pretext of needing it for growing soybeans.[59] Bauxite did, however, form a large part of Guyana's primary commodities being exported around the time of independence in the 1960s.[60]

The Sovereign Land Claim

Sovereign (de)colonial mentality continued in specific relation to the forests in the form of the definition of national forests and encompassed those rules, regulations, and organizations guiding forest usage. Through executive authority enshrined in the Constitution of Guyana and control over the legitimate use of force, the independent state government of Guyana exercises sovereign (de)colonial mentality over the forests within the territory. The Mining Act, Amerindian Act 2006, and Environmental Protection Act 2006 have implications for REDD+ in Guyana. The Mining Act and its associated regulations govern mining and outline the process of administering permits. With gold mining forming the major threat to forests in Guyana,[61] its observance and execution has major implications for REDD+. The mining law, enforced by the Guyana Geology and Mines Commission (GGMC), claims all minerals throughout the country as property of the state, even when the land is under the tutelage of indigenous communities and private persons. The Amerindian Act allows for the demarcation and ownership of significant sections of land that could be used for conservation activities and climate change mitigation. The Amerindian Act allows Amerindian Village Councils to lease communal lands for mining, forestry, hunting, and residential purposes. Amerindians collectively have tutelage over 13.9 percent of Guyana's landmass, with Village Councils designated as the governing bodies of these lands and vested with the responsibility for their conservation. The Environmental Protection Act 2006 enables the "management, conservation, protection and improvement of the environment, the prevention or control of pollution, the assessment of the impact of economic development on the environment and the sustainable use of natural resources,"[62] and it designates the Environmental Protection Agency (EPA) as its governing body. Other relevant pieces of legislation include the land law that outlines the process for managing state land; the Iwokrama Act of 1996 that provides for the sustainable utilization and management of the Iwokrama International Centre for Rainforest Conservation and Development of approximately 360,000 hectares for research; and infrastructure policies and international agreements impacting forest use, such as the United Nations Framework Convention on Climate Change (UNFCCC) and the UN Declaration on the Rights of Indigenous Peoples.[63]

Together, these arrangements form the legal base for REDD+'s functioning in Guyana. Parts of this legal framework for utilizing the forests were revised in response to international interest in climate change mitigation through carbon sequestration. The impetus to address climate change on the international level motivated successive independent governments to consider the role of forest conservation in economic development.[64] These activities take place within an organized policy and legislative framework where certain governmental bodies are designated for working with REDD+. In these ways, sovereign (de)colonial mentality came to be continuously expressed, although constantly in flux, throughout the history of Guyana. Through sovereign (de)colonial mentality established during the colonial period, independent state governments were designated as the primary actors for forest governance and REDD+ readiness over a particular territory.

Access to land was and still is a pivotal factor in the economic fortunes of the different ethnic groups in society. Forests, in particular, had been taken over as an area to be managed by the state. The legacy of this land claim by the colonial and independent governments of Guyana is entrenched and legalized in the constitution where it is written that "the Minister may by order declare any area of State land to be a State Forest and may, from time to time, vary or revoke such order."[65] Further, article 36 of the 1980 constitution establishes the state's emphasis on rational use of its natural resources through an institutionalized structure including the Guyana Forestry Commission (GFC), Guyana Lands and Surveys Commission (GLSC), Guyana Geology and Mines Commission (GGMC), Ministry of Amerindian Affairs (MOAA), Ministry of Agriculture, Ministry of Local Government, and Guyana Environmental Protection Agency (EPA), all responsible for different land use practices often on the same plot of land.[66]

The forests were also categorized and classified, with the GFC declaring that while 87 percent of the country's 21 million hectares of land resource is covered by forests, 18.3 million hectares are forested, of which 12.8 million hectares are state forest administered by the Guyana Forestry Commission. These forests have too been duly categorized as swamp, seasonal, and dry forests in different parts of the country.[67] This zoning of the forests for state-sanctioned uses across different zones sets the conditions for the competing land claims of those who had been living in and utilizing the forests for centuries.

Linked Fortunes but Different Destinies

Politics in both Guyana and Suriname is largely based on support from various ethnic groups, reflecting divisions that have manifested themselves in violence between groups in both countries at different points in their histories.[68] Reflecting on the colonial legacy in modern-day Guyana, an advisor of the PPP-led government on climate change–related matters stated:

> The fact that we are a plural society here is an example of that, where at the end of slavery, the freed slaves, of course, were unwilling to go back to the plantations. They were given a hard time. Even though they bought land, they were not given drainage and irrigation. They were not allowed to function in their villages. Then, the indentured laborers were brought in: Chinese, Indians, and Portuguese, and just lumped together without any thought as to their future in terms of education, the coexistence in terms of culture, religion, and all of those things. So we very much have been left to fend for ourselves, and as you see the bad examples of tribal wars in Africa, and even in the post-independence period, people still trying to come to grips with what exactly is this independence.[69]

There, in the above quotation, is presented a clear reflection of how these countries continue to grapple with their identities with the outside world, and amongst themselves. Succinctly encapsulating the need for decolonization, the advisor went on to state:

> If you are mentally still subjugated, subordinated, whether it is to World Trade or the academic education, you are only allowed to pursue certain vocations, and even your knowledge of your own country. I knew more about British history and British geography than I knew about the Caribbean and my own country. When you consider that we are only 48 years old as an independent country,[70] then it is too short a period to really expect an overturning of those aspects of colonialism.[71]

In general, colonial histories continue to shape independent Guyana as the multiethnic population negotiates racial strife and discord, which is often expressed around environmental use practices and development objectives.

BECOMING SOVEREIGN—SURINAME

While the history of indigenous people prior to colonization in Suriname is similarly not well known,[72] it is asserted that Arawak and Carib groups were at war when they were confronted by European colonizers in the early 1600s,[73] and in Suriname indigenous people had also relied on the subsistence economy to meet their daily needs.[74]

The Colonial Sovereign in Suriname

The eventual colonization of the area that became Suriname took a path similar to that of Guyana, but with a different outcome, in that Suriname gained independence from the Netherlands, despite intervals of British rule. Events in Suriname also followed a familiar pattern of divide and rule. The first successful establishment of a European settlement in Dutch Guiana, the area now known as Suriname, was in 1650 by British planters who eventually began to establish plantations through the use of slave labor on the coast.[75] The colony changed hands from British to Dutch rule in 1667.[76]

The planters in Suriname also exploited "red slaves," as indigenous slaves taken by the Caribs were known. However, due to the Dutch presence on the coast, the indigenous populations of Suriname withdrew into the country's forested areas. The only remaining contact between the Europeans and Amerindians took the form of trade in items such as weapons and cloth.[77] This withdrawal strengthened Suriname's social separation between the coastal and forested zones.[78] Its legacy can be seen today in the residence patterns of most indigenous communities, which remain deep in the forests away from the more populated coastlands.

The withdrawal of indigenous people from the coast was not entirely without resistance, however. In 1678, the Caribs went to war with the Dutch colonizers, who were at the time still weak. In 1684, a peace treaty was signed between the Dutch and the Amerindians. As a result, indigenous people were again free to live in Suriname's interior with livelihoods based on subsistence economies as they had done prior to European arrival. Suriname fell back under British rule in 1799. After subsequent brief periods of Dutch and British control, it again came under the control of the Dutch for the final time in 1816.[79] Through this series of events, the external borders of modern and independent Suriname were established. These borders, though contested, lie between Suriname and Guyana. They underpin the state demarcations used for national REDD+ efforts.

Like British Guiana, the colony of Dutch Guiana also depended on the importation of large numbers of people from other continents to work on the plantations set up by Europeans on the coast. The importation of enslaved Africans, in particular, eventually resulted in communities of runaway slaves, now called maroons, taking up residence in the forests and adopting some indigenous ways of life.[80] The plural society that emerged in Suriname, so categorized due to the importance of ethnicity in the daily life of people,[81] was based on a small white plantation class, a relatively large number of enslaved Africans, and a creole section of the population composed of mixed African and white descendants. The maroons went on to form villages in Suriname's forested interior, while the coast remained divided along racial and color lines.[82] In Suriname, like Guyana, the physical separation of the forests from the agricultural coastal areas serves as a clear reminder of the colonial period. It can also be seen socially in how forested communities are often viewed and depicted in policy and development documents as traditional, vulnerable, and backwards, in need of support from their more developed counterparts.

The Independent Sovereign of Suriname

After independence, creoles came to dominate Surinamese politics. Creoles are the descendants of enslaved Africans who established themselves in the city after emancipation. The maroon communities in the forests were reluctant to acquiesce to independence because they considered themselves already mostly independent. They had built up an amicable relationship with the colonizers with whom their forefathers had made peace in the 1760s. Instead, maroons feared domination by creoles in the city, who at times claimed to be able to represent maroon interests,[83] drawing on a perceived commonality based on shared skin color. Suriname's post-independence period was defined, however, by internal civil strife that brought about the fall of the economy and the severing of Dutch development aid. These two connected events left indelible imprints on the country's future.

Suriname benefitted from a "golden handshake" in the form of a large development aid grant from the Netherlands. Through this grant, the Dutch were able to retain substantial control over the Surinamese economy and its internal affairs.[84] The grant, formally called the Multi-Annual Development Program (MADP), was to be disbursed over ten to fifteen years commencing in 1975–76. It should have amounted to some 2.2 billion Surinamese guilders being spent on a population of four hundred thousand people. This was unlike the arrangement of any other colonial power at the time of independence.[85] Using the per capita income of Suriname at the time, the grant reflected an increase of between 300 and 600 percent of the national income. The Dutch had among their hopes that the grant's disbursement would reduce possible immigration to the Netherlands before independence was formalized, by making Suriname a more desirable place to live.

Through the grant, the Dutch sought to encourage diversification of the economy and to promote self-sufficiency and overall development within Suriname.[86] Apoera, a forested majority-indigenous community near the Corentyne River that separates Suriname from Guyana, was flagged to receive a significant portion of the grant, in aim of having it develop as a new city that could produce hydropower for the bauxite industry.[87] However, the program did not achieve its aims and ran into difficulty in terms of disbursement because Suriname was ill equipped to absorb these large sums.

A bloodless military coup took place in 1980 that was driven, in part, by dissatisfaction with the way that the grant was being disbursed. The new military government prepared a new plan and convinced the Dutch government to use a minimum of 500 million Surinamese guilders to fund it. This was, however, the beginning of threats from the Netherlands to stop the grant. During 1980 and 1982, numerous countercoup attempts were made, by members of the economic and political elite that had been removed from power, and quite likely also by the former colonial power, which did not approve of military rule in Suriname.[88] This situation culminated in the murder of sixteen political opponents by the Surinamese military in December 1982. The Netherlands swiftly and indefinitely suspended the aid agreement. The Dutch then worked to isolate Suriname further in the international community due to its stated desire of the Netherlands to see human rights and the principles of democracy reinstated in the country.

Internal strife driven by Ronnie Brunswijk further complicated the matter.[89,90] Ronnie Brunswijk, the leader of the Suriname Liberation Army, functioned as a sort of Robin Hood in Suriname society, sharing the spoils of his illegal activities with maroon communities in Moengo. This spurred retaliation by the Surinamese army, which carried out violent atrocities against the maroons who supported Brunswijk. This resulted in a civil war that took place between 1986 and 1992.[91] The war plus the fall of the price of bauxite on the international market devastated the Surinamese economy. Debt and inflation soared in the late 1980s. In the early 1990s, as a result of the war, education stopped in the maroon areas. As a result, the young male population turned largely to gold mining as a source of income, a shift with profound later implications for forest conservation and the feasibility of REDD+.

Democracy returned to Suriname with a referendum in September 1987, an election in November 1987, and the handover of power to the new government in February 1988. However, the Dutch government, instead of resuming aid, devised new methods of providing funding that featured the involvement of international financial organizations in restructuring the Surinamese economy. During the period of internal conflict, garimpeiros moved over to Suriname to work after having lost their land usage rights to big business in nearby Amapá, Brazil.[92] They often collaborated with maroon gold miners and brought along technologies that were new to gold mining in Suriname at the time. Some maroon communities in the gold mining areas created a revenue system by informally taxing Brazilian gold miners a bit more highly than maroon gold miners. Despite this, Brazilian gold miners are known to have a relatively tension-free relationship with Surinamese maroons and the wider Surinamese society.[93] Since independence, however, Suriname's presence on the global stage has remained minuscule. This is due, in part, to the tendency of the international community to leave the country out of global deliberations because it is not a part of Spanish-speaking Latin America or Portuguese-speaking Brazil or squarely a part of the English-speaking Caribbean.

Racialized Identities and Environments

The internal structure of Surinamese society continues to be influenced by tensions between its different ethnic groups,[94] but the situation remains relatively peaceful. Racialized identities are evident in how these groups continue to be aware of their historical connections. Maroons, for example, spoke freely of their living memories. When asked about their living knowledge of the way they were brought to Suriname, one person said, "Our grandparents came out of Africa, and they lived there for around 100 years. . . . It was a nightmare for us, and it is traumatic for us. We will never forget it."[95] When asked about the way they make a living now that they are settled in Suriname, one respondent called Ben said, "Wood. . . . We are living from the rainforests, we use everything. Also, the gold mining in the woods. . . . We are dependent on the forests."[96] Maroon communities subsist on farms and trade with the capital city. Increasingly, they are turning to small-scale gold mining as a source of income. Ben continued: "That's the problem over here, we are still living as slaves."[97] His reference to living as slaves is grounded in his experience of being chased off mining concessions allocated to multinational large-scale mining companies, impeding his small-scale gold mining activities.

Different stakeholders of REDD+ in Suriname describe the respective relationships between maroons and indigenous groups with the forests in which they live as being different. LMR, an indigenous community member, stated:

> Yes, the indigenous people are not interested in mining because they have a special bond with nature. . . . I think the bond [the maroons] have is "we respect the forest because the forest can kill us, and it can give us food." It is a very strange relationship they have with the forest, with the jungle. The indigenous people don't look at the forest in that way. They always speak of the forests like it is a person. It has feelings that can speak to you. It is totally

> different compared to what the maroon people do. They live in the forests, but it is like the forest is not, they don't see the forests as a person that has feelings.[98]

According to respondents who share this view, these different relations to nature have an impact on the views of these two groups of communities on land tenure. The respondents highlight that maroon societies tend to be more individualist in orientation, while indigenous communities are more collective. A respondent posited that "it could be a historical thing because they [maroons] sort of were forced into the interior. They weren't there before, and they learned from the indigenous people how to live and what to do."[99] I use these interview excerpts to illustrate how different ethnicities within Suriname have taken up certain socioeconomic positions.[100] These positions inform their relationship to the land and forests.[101] They have taken up certain spatial positions that have led to "ethnic labour specialization, economic stratification, and spatial concentration and segregation."[102] My goal here is to highlight how the shape-shifting, racialized subject, informed by colonial histories, remains an ever-present actor informing development outcomes and relations with the natural environment.

As previously noted, both Guyana and Suriname played key roles in supplying bauxite to the United States during the Second World War. American interest in bauxite in Suriname, however, stretched back to 1917 when Alcoa, which also operated in Guyana, established a branch in Suriname. Suriname's reliance on bauxite continued strongly even into the late 1950s, as its bauxite was needed for the supply of aluminum to support the US war effort in Korea.[103] By 1975, bauxite production was of such importance to Suriname that plans were made to develop a new city around its refinement, along with a railway and road system that would facilitate its extraction and transport. As previously noted, this new city was to be located at the site of Apoera.[104] Although these plans never came to fruition, Suriname's reliance on bauxite mining clashed with the land claims of some communities in the forests.[105]

Notably, one large infrastructure project related to bauxite mining that did materialize continues to affect the well-being of several forest communities within Suriname. The Afobaka dam was constructed to provide hydropower for a bauxite refinery. The construction of the Afobaka dam

Figure 6. Ben pointing to a house given to the relocated maroons.

forced the relocation of maroon communities from the surrounding area. As described in the introduction of this book, the construction of the dam and its accompanying reservoir in the Brokopondo district of Suriname forced approximately six thousand Saramaka and Ndjuka maroons off their land in 1963–64 during the administration of Dutch colonial government. Some six hundred square miles, amounting to almost half the territory claimed by Saramaka maroons, were flooded. The communities were paid the equivalent of US$3 in compensation and were not assigned land rights in the areas to which they were relocated.[106]

Ben, who was previously quoted talking about the historical memory of slavery as a maroon today, described his experience with the construction of the Afobaka dam. He said:

> I was born in the District of Suriname, Brokopondo, and when I was little, I had moved from Stuwmeer [Stuwdam] to here [Brownsweg], when I was four years old. [We moved] because of the power, because we were

> living near the river, African slaves, then we had to move because they had to dam the Stuwmeer, but it was really sad. That's why we had to move here. . . . We had to move. There was no choice. We had to. Otherwise we would die.

Suriname's gold rush began in the late 1860s. Initially, the labor force on the small gold mines was made up of creoles, with Amerindians and maroons seeming little interested in the endeavor. The maroons used their knowledge of the rivers and jungle to provide transport for equipment and people to and from the gold mines. The extraction and sale of gold continued to be big business in Suriname until World War I, when demand slumped.[107] This void in economic earnings was filled partly by the bauxite industry, which as previously described, gained economic value as demand for aluminum increased to supply the demands of war.

The experience of being forced from their lands brought about loss of human life in the communities. The devastation did not stop there, however. Community members in Nieuw Koffiekamp, one of the areas to which the displaced Saramaka maroons were relocated, faced a second relocation in 1995 after their lands were granted as a concession to a Canadian gold mining company. The community was not made aware of the granting of the concession until they were surrounded by trucks and armed guards who restricted their subsistence and small-scale gold mining activities and intimidated them by firing live ammunition to keep them from the areas in which the Canadian company was working.[108]

Nonetheless, these relocated maroons now engage in gold mining as their major source of income. Gold mining has become the main source of income for many Saramaka maroons now resident in the Brownsweg area. Consequently, the government of independent Suriname faces the difficult challenge of reducing small-scale gold mining there. Maroons in the area are vilified, along with the small-scale Brazilian miners, for being the main source of deforestation in the country. There is, however, a desperate need in Suriname for improved management of gold mining, which is known to be a lawless activity, invested in by businesses in the capital and by drug dealers.[109] The government of Suriname began to make efforts to coordinate and manage the devastating environmental and social effects of gold mining in the country through the Gold Sector Planning Commission (OGS in Dutch).

Notable also is the consideration that, in Suriname, a large portion of national economic earning activity is directed towards boosting GDP and maintaining the large public sector, accounting for as much as 60 percent of employment.[110] This income is maintained in part by the extractive industries. According to Bruijne and Schalkwijk, this is because political parties in Suriname tend to use the public sector as an avenue towards allowing access to the state's resources by their popular or ethnic bases.[111] These scholars further explained the distribution of employment benefits across a coalition government, stating:

> In the past, government ministries and other public offices were distributed among coalition partners to ensure "equal access," between different political parties and thus their ethnic base, to public resources, such as civil service jobs, housing, land, loans, and permits. . . . Ethnic identity and ethnic boundaries have thus not been eliminated by the cultural and political context, but are sometimes exacerbated by them, though members of the ethnic groups apply them in a flexible rather than a rigid way.[112]

As one of my respondents in Suriname explained:

> I think there is a huge public sector machine, if you wish, like way, way, way too many public workers. . . . Every time there is a new government, other people become non-active, and then there are new people being hired. You can't fire the other ones because you know, they are public workers. There is a huge pool of workers being paid by the government and they all have, you know, not all, but people at a higher level, they have cars, they can get free gas, and then, people, persons who have been in the government in higher positions, they have basically benefits and pensions for life, so even if they have just been serving in the national assembly for a couple of years, they are kind of covered for the rest of their lives. So, it's such a huge amount of money that is being used by that. . . . All the cars being driven by government workers at all levels. You wonder why they need to be so expensive.[113]

This burden impacts the Surinamese state's reliance on extractive industries for earning an income. In addition to extractive activity, Suriname's economy is dependent on aid, although it also covertly benefits from illegal gold mining, narcotics trading, and remittances from Surinamese living in the Netherlands.[114] Some foreign exchange also comes from the country's small onshore oil industry.

Suriname's ethnic composition is also complemented by a significant number of people of Chinese descent. Suriname was the recipient of a large influx of Chinese nationals, during the colonial period, who served as indentured laborers on the coastal plantations. However, up to 2010, Chinese engagement with Suriname was minimal.[115] Thereafter, the Chinese population grew from a relatively limited number of largely affluent people to larger numbers of poorer immigrants. The Chinese-identifying portion of the population increased from 3 to 8 percent. Many of the immigrants are in Suriname illegally. They are either smuggled into the country or overstaying visas previously allotted to them. Older generations of Chinese in Suriname help to establish and to make room for newer generations of migrants and often act as local counterparts for projects based in mainland China.[116]

The recent influx of Chinese nationals has not been devoid of tension. Contrary to European and American tensions with Suriname due to the colorful past of its former President, Desi Bouterse, China's relationship with Suriname is nonjudgmental and welcoming, with Suriname being one of the first South American countries to establish relations with China. Suriname's relationship with China, however, remains both similar to China's resource extractive relationship with other countries, and different in that there is now a large Chinese population within Suriname that occasionally conflicts with other ethnic groups.

Chinese companies are deeply involved in infrastructure projects within Suriname, and the award of some projects has been highly contentious. Many projects are on stream, but compared with some of its neighboring countries, Suriname has remained less deeply dependent on Chinese loans for the execution of these projects. China's biggest commercial interest is in lumber. Even though they are not permitted to own timber concessions in Suriname, Chinese companies have been able to buy the concessions of others operating there, as mentioned in chapter 1. Gold mining is also potentially interesting to Chinese companies, but Suriname's interior is currently beset by some forty thousand individual Brazilian miners looking to extract the gold that was unexploited during the last century of mining in Suriname.[117] The tensions between some ethnic groups in Suriname and the growing Chinese presence were palpable during my data collection period. They formed a growing undercurrent that bubbled occasionally to the surface when, for example, a Surinamese creole person remarked, while discussing service options for changing a car tire, that the vulcanizing shops are still quite expensive because "the Chinese haven't set their sights on that yet."[118]

The ethnically mixed but mostly indigenous community of Apoera also demonstrated a certain resignation to the Chinese lumber firm operating in their backyards. The maroons shared similar concerns in their expressions of annoyance with the Chinese working in forests near their communities. The Chinese were said to be using dogs to protect their outposts from community members. Chinese shops were also easily spotted in maroon villages along the main road that runs through Brownsweg, with at least four Chinese supermarkets and one Chinese restaurant present in the small grouping of communities. Surinamese citizens reported feeling as though the Chinese are taking advantage of every conceivable economic opportunity. They also reported viewing the influx of Chinese into the country as having the potential to put good environmental stewardship at risk through poor management practices, sometimes with covert Surinamese government support.

The Sovereign Land Claim

In 1975, the constitution of an independent Suriname came into force. As quoted in the opening pages of this chapter, it emphasized that the state has the "inalienable right to take complete possession of the natural resources in order to apply them to the needs of the economic, social and cultural development of Suriname."[119] Unless private ownership of land can be proven, land and the forest covering it are considered to be under the jurisdiction of the state. However, in the case of the forested indigenous and maroon communities, such evidence that land is individually held does not exist.

The national definition of *forest* is included in Suriname's 1992 Forest Management Act. However, the definition focuses on production forest, making it unsuitable for monitoring and conserving forest cover. Therefore, Suriname's REDD+ readiness and implementation process, dependent on a different type of forest "use," requires the adoption of a new definition for national forests.[120] Still, a collection of other laws provides for the conservation of carbon stocks and the sustainable management of natural resources. These are the 1954 Nature Preservation Act establishing administrative arrangements for maintaining natural monuments, the 1973 National Planning Act supporting national and regional planning on issues related to land use policy, the Forest Management Act of 1992 establishing a legal framework for forest management and sustainable utilization of forest resources, and the 1998 governmental decree on nature protection establishing the Central Suriname Nature Reserve. An Environmental Framework Act is being prepared to regulate pollution, waste, and other environmental impacts. Otherwise, the national development policy, climate change strategy, and National Climate Action Plan contribute to mitigation and adaptation.[121]

However, some less impactful regulations do recognize customary resource rights, such as the 1992 Forestry Act and the 1998 Nature Protection Resolution. After the conclusion of the aforementioned civil war that took place largely in the forests of Suriname between 1986 and 1992, the Suriname government committed to resolving the claim of forest communities for land rights but has not done so, despite having signed legally binding national-level documents and having ratified several international treaties that commit them to respecting indigenous rights.[122] In addition, mining and logging concessions that had been granted by the government continue to exist in traditional lands. National parks, also subsequently formed, impose restrictions on traditional practices.[123]

Linked Fortunes but Different Destinies

The extractive industries formed a core part of the economic development of Guyana and Suriname in the run-up to and period after independence. These economic earning activities were spurred largely by demands and events taking place in the international arena, as Guyana and Suriname's natural resources became coveted, in fits and starts, for their desirability in the international markets. However, the pursuit of these activities was as racialized in nature as were the different ethnic groups brought to labor in the colonies of Guyana and Suriname. These economic activities have become part and parcel of the culture of certain ethnic groups in these two countries. Therefore, international environmental policies seeking to protect the forests of Guyana and Suriname can scarcely avoid these deeply entrenched histories that continue to shape the fortunes of both countries.

The major difference between the ethnic profiles of Guyana and Suriname lies in the presence of significant minority populations of Indonesians and maroons in Suriname. In comparison to Guyana, however, Surinamese society shows less dominance of a particular ethnic group, as is evident from the fact that the country is often governed by an array of small parties that have formed coalitions, and from the more ethnically integrated nature of their capital city. While racial tensions continue to exist in Suriname, they are less obvious and palpable than in neighboring Guyana.

In both societies, however, the plantation economy and the features of slavery and indentureship continue to influence the societal structure, ethnic composition,[124] and environmental use practices, as evidenced by the living patterns and economic earning activities of different ethnic groups. Social separations in Suriname are evident through language, with different languages being spoken by different groups. The economic and racial separations in Suriname are somewhat limited on the coast but continue to be impactful in the forests, with some maroon communities primarily engaged in mining, and indigenous communities generally working to maintain their subsistence practices. However, both countries have resource-based economies that make them vulnerable to the vagaries of the international market. This situation remains, despite attempts in Suriname to diversify the economy after the turn of the nineteenth century.

Politically, Suriname's politics is characterized by a number of coalitions and the fractious nature of the population, resulting in a situation where it is difficult for a single party to win an outright majority.[125] Guyana's independence of 1966, on the other hand, took place in the foreground of the Cold War, which saw several foreign interventions in its politics, and these served to uphold political divisions according to race. These different histories certainly lead to different present-day political circumstances and continue to influence the way present-day Guyana and Suriname view themselves in relation to the rest of the world.

Despite many shared circumstances, a rather harsh separation exists between the two countries, representing a leftover relic of the colonial pasts of Guyana and Suriname. These two are engaged in a protracted territorial dispute that directly challenges carbon accounting methods because both countries claim the same territory and its forests as their own. The contested area is referred to as the New River Triangle in Guyana and as Tigri in Suriname. Both countries include this swathe of forests and its carbon sequestration work in their REDD+ proposals. While this issue was not highlighted during public REDD+ discussions in either country, the border dispute is a continuous sore point between Guyana and Suriname resulting in the occasional skirmish between the two armies. This dispute represents one of the most easily identifiable examples of how colonial histories and their resultant border politics continue to impact the preparation for and implementation of REDD+.

RACIALIZED ENVIRONMENTS AND THE ANTHROPOCENE

This chapter showed that Guyana and Suriname have much in common, including multiethnic populations established through colonialism, a shared landmass, overlapping Dutch and British colonial histories, and densely populated coasts. They are also highly vulnerable to climate change—both the physical and governance-related aspects of which tend to overlap with colonially rooted, labor-inflected, racial population distribution patterns.[126] Consequently, in their post-independence periods, these two states must grapple with rapidly intensifying climate change while featuring multiethnic populations where whiteness has been largely relegated to global imaginaries of global wealth and development interventions.

This chapter was written in the awareness that much of the research that examines the relationship between processes of race and climate change is situated within debates on the Anthropocene. The Anthropocene was conceptualized by meteorologist and atmospheric chemist Paul Crutzen and biologist Eugene F. Stoermer, who argued that the Earth is in a new geological age in which human beings and their activities are the defining factor in influencing nature and the climate.[127] The Anthropocene concept recognizes the myriad ways in which human beings have altered the state of the planet, amongst which climate change is but one. The concept has been fiercely debated, however. It has become a point of contentious debate on a variety of pressing global concerns that include considerations of how to conserve nature in the face of environmental decline,[128] processes of racialization in terms of who is represented or not within the designation of "Anthropos" as the actor responsible,[129] the place of capitalism as mediating force between humans and nature,[130] and the hierarchies of knowledge implicit in its conceptualization.[131]

Much of the literature critical of the Anthropocene challenges the concept's underlying idea of an undifferentiated humanity in ways that upend ascriptions of novelty to the current epoch. Critics argue that it is a racial construct that extends to the entire globe a certain Western conception of nature, erasing as it does so the racial injustices and colonial histories of extraction and exploitation that have implicated most of the world's population in its emergence only by force.[132] Pointing to actors, other than a broadly conceptualized "Anthropos," who are instrumental in ushering the world into this new alleged epoch,[133] critics also attribute primary responsibility for alternative "-cenes" to varied phenomena through renaming the epoch with neologisms such as "the Chthulucene" and "the Plantationocene."[134]

In view of Anthropocentric debates on conservation, political ecologists and market-based conservation skeptics Bram Büscher and Robert Fletcher argue that conservation tends to rely on problematic separations between nature and culture, separations they see as fundamental to both capitalism's functioning and conservation's inability to stem ongoing and dramatic ecological decline.[135,136] For them, the Anthropocene concept "conceals the reality that different groups of people have vastly different environmental impacts behind the image of a generalized 'humanity.'"[137] They argue that "conservation and capitalism have intrinsically co-produced each other, and hence the separation between nature and culture is foundational to both."[138] Challenging the problematization of "Anthropos" in the Anthropocene in favor of "Capital," they write that "humans cannot overcome the 'age of humans'" but can and must "overcome the 'age of capital.'"[139]

However, many of these critiques, even those that challenge the concept on the grounds that it overlooks racial difference in its attribution of responsibility, have largely been expressed through a global classification of human beings racialized along a continuum from whiteness to blackness. While this literature has established that climate change is having disproportionately negative effects on people in the Global South and on nonwhite populations in the Global North, the manner in which race and processes of racialization play out within these large-scale racial groupings in efforts to represent and to combat climate change is less commonly examined. The color spectrum leaves little room for understanding the situated environmental and social nature of racialized, colonially inflected environmental relations in multiethnic societies where whiteness is no longer an immediate and significant local limit, such as those evident in Guyana and Suriname, where, simply put, the colonizer in fleshy form has gone home.

This neglect overlooks the way that efforts to combat climate change can make some groups within all-consuming categories like blackness more vulnerable than others, both to climate change and to efforts to remedy it. Hence, the intersection between race, climate change vulnerability, and governance comes to be portrayed in these critiques according to the too-simple, color-coded, global continuum of race. Significant complexity lies within those groups of people in areas that are often represented as somewhat devoid of agency in these discussions. In other words, though "Anthropos" of the Anthropocene is white,[140] research that investigates the way that climate change, along with efforts to mitigate it, differentially impacts and is impacted by people within nonwhite racial categories is also warranted.[141] Instead, I argue that although a relationship to whiteness was instrumental in the creation of Guyana and Suriname, and continues to shape their engagement in the global political economy, it is no longer the primary means through which people continue to be racialized in addressing and being affected by climate change. Instead, their racialization remains pinned to the environment.

As I observed elsewhere,[142] Guyanese historian, academic, and activist Walter Rodney commenced his book *A History of the Guyanese Working People, 1881–1905* with an account of the massive effort through which the narrow coastland of Guyana, on which the vast majority of its population now resides, was reclaimed from the sea.[143] He wrote: "An enduring Dutch and European contribution to the technology of Guyanese coastal agriculture was undeniable. Yet one must guard against the mystification implicit in the assertion that it was the Europeans who built the dams and dug the canals."[144] Instead, it was enslaved people from Africa and indentured servants from India who "had to face up to the steady work diet of mud and water in the maintenance of dams and the cleaning of trenches."[145]

A reading of Rodney's account might support two-part, exclusively social interpretations of how the white colonizer played a central role in the historical construction of race by bringing different groups of differentially exploited people to the then colony to labor in support of capitalist development.[146] However, his account can also be read differently—as indicative of a tense and tumultuous three-part relationship between (first) the environment in the form of mud and water, (second) white Europeans, and (third) black and brown workers from Africa and India through whose collective actions the climate-vulnerable coastland emerged and came to prominence. These European-directed, Asian- and African-executed battles against the environment informed not only the racialized subjectivities and identities of different groups of people in relation to each other, but also their racialized relationships with the environment. Hence, Rodney's account highlights how what can be now described as the natural environment, as discursively powerful and neutral whiteness, and as base, impure, and exploitable blackness were all abstracted and co-constituted in small part through the creation of the climate-vulnerable coastland.[147,148]

The work of Patrick Wolfe supports my distrust of color-coded interpretations of race by providing a significant challenge to the internal coherence of each constituent color.[149] He commenced with an observation of how these color-coded interpretations were used by some American students in Australia who were taking his class on aboriginal history. He noted that the American students were interested in seeing how people they saw as black fared in Australia and implicitly drew connections between the experiences of African Americans in the United States and those of indigenous groups in Australia, based on their shared "blackness." Wolfe responded by explaining that the groups in Australia to whom they were referring, while also often referred to as black in Australia, were in fact indigenous people, with their closest counterparts in the United States being what Americans there would refer to as Native Americans or, according to color codes, as red. The few indigenous Australian students present participated in these discussions by demonstrating their strongest expressions of kinship not towards African Americans, as might be expected in color-based categorizations, but towards indigenous communities in the United States with whom they shared similar histories of invasion, loss of land, precarity, and resistance. This distinction provides an important correction. Given that most of the literature that examines the overlap between race and climate change can be found in the literature critical of the Anthropocene, which tends to rely on global color-coded interpretations of race,[150] my arguments offer an analysis of the intersection between race and climate change that similarly challenges the homogeneity of blackness as a category on which to exclusively base climate change vulnerability and mitigation interventions. In so doing, this book makes space for racialized relations with the environment to come to the fore as indicators of how colonial histories continue to resist and limit neoliberal conservation methods like REDD+ in the global fight against climate change.

CONCLUSION

The histories of Guyana and Suriname, especially the deeply rooted tradition of using their resources and people as stores of value to be exploited, have had a profound impact on how different groups and cultures in these countries relate to the environment. In this chapter, I traced not only the racialized relations with the natural environment, but the legacy of the colonial and state claim on the forests. These state claims effectively set the groundwork for current forest use practices and the concomitant need for customary rights to be recognized as an exception to these rules based on ethnic categorizations. However, forest communities continue to challenge the state by asserting usage of the land for generations prior to the formation of the state. In so doing, they resist sovereign means of governing the forests. After all, the process of politically demarcating the forests and of establishing racialized relations of its use was sovereign in nature, based on the colonial and then independent sovereign claim of what was previously indigenous territory followed by its subsequent classification, categorization, and management through laws and regulations.[151] To a large extent, the process of managing the forests of both countries continues in this fashion, supporting market operations and the integration of these forests in the global spread of carbon markets as an adjudicator of climate change and development concerns.

Despite broad areas of commonality, there remain clear differences between the two countries. Yet, both independent governments continue to be haunted by the presence of the colonially rooted racialized subject, which relates to nature in particular ways. REDD+ is therefore made to interact with this racialized subject across both countries. This racialized nature of the subject of REDD+ governance takes on harsher tones in forested communities, where less societal integration with other groups took place over the centuries, alongside less geographic movement of people and less communication with the coast. Thus, in both countries, forested communities depict a harsher tone of "racialized-ness" than on the coasts.

Differences and similarities notwithstanding, the point being made here is that a racialized subject and associated environmental use practices exists as a legacy of colonial histories with which REDD+, as a market-based incentive, must interact. The political contestation related to land rights in both countries represents a gap between state control over land, and the process of allocating satisfactory and effective customary rights that has become a major sore point for the implementation of REDD+. I identify this sore point not to position concerns for land rights against arguments in support of REDD+ implementation. Instead, I point to the contestation around land rights in both countries as a means of showing how REDD+ becomes mired in preexisting, colonially rooted conflicts around resource access, rights to the land, and ethnic and political conflicts in the areas in which it has been implemented.

Thus, REDD+, as a novel instrument developed for altering forest use practices, is interacting and negotiating with these different subjectivities, which are by no means uniform. The circumstances of the maroons of Brownsweg, for example, in this capitalism-fueled era of climate change further represent the increasing inescapability of the reach of capital, subjectively experienced in their daily lives and views of themselves, and physically evident in the degradation of their changing environments. This reach is extended through and in response to national neoliberal development trajectories, degradation of the environment being experienced around the world, and the neoliberal conservation methods developed to address this degradation.

Consequently, I posit that decolonizing environmental governance in the Amazonian Guiana Shield depends first and foremost on recognizing and dismantling the colonial histories of oppression, forced relocation, and violence imbricated in the now-independent and sovereign Guyanese and Surinamese states. This recognition should be accompanied by an awareness of how the identities of different groups of people continue to negotiate the racialized societal structure that came about alongside the state's emergence. These identities also continue to shape their natural resource use practices today. Decolonized environmental governance should recognize these path limitations in support of decentering and cutting off the sovereign head of the state governments.[152] Once the state as sovereign has been beheaded, by which I mean removed from the position of default actor of top-down government, different approaches of historical forest governance may come into view. I analyze some of these approaches in the chapters that follow. Next, I put forward a case for the second step in the journey towards decolonizing environmental governance—decentering the market logic on which the independent sovereign states of Guyana and Suriname have come to increasingly rely, and which aligns unsurprisingly well with the market-based incentivizing logic of forest conservation through REDD+.

3 Decenter Markets

ANNAI

At the foot of the Pakaraima mountain range lies a group of indigenous satellite communities called Annai. Annai is a small village hub of conservation, development, and REDD+ activities. En route to Annai, Guyana's Caribbean-ness begins to fade as the feel of a South American landscape takes over. Close to the southern border with Brazil, Annai is accessible by either a short flight from Georgetown or a tedious seventeen-hour drive along a dirt trail through the forests, weather permitting. Imperceptible upon your arrival, however, is the fact Annai represents a peculiar confluence of circumstances. It acts as a hub of knowledge, science, and forest monitoring, and as an example to other indigenous communities in the Guiana Shield seeking to develop sustainably. Annai is well-known for its focus on sustainable development and for the way it harnesses its natural resources for the development of its people. Moreover, Annai is located outside the Iwokrama conservation area, a reserve of 371,000 hectares of forests set aside by the state government of Guyana to be sustainably managed to the ecological, economic, and social benefit of the people of Guyana and the world. The Iwokrama conservation area draws on inter-

Figure 7. Solar panels in Annai, North Rupununi, adjacent to one of its buildings displaying the People's Progressive Party (PPP) flag (Collins, 2014).

national partners and actors to develop innovative ways of managing forests. Hence, the villages within and around the conservation area benefit tremendously from its proximity to the conservation area in terms of knowledge sharing and support.

Annai, with a population of some 523 persons, acts as a base for the North Rupununi District Development Board (NRDDB). The group of communities hosts several internationally and locally funded projects and nongovernmental organizations working towards sustainable development. It relies quite heavily on technological means of managing natural resources. The World Wildlife Fund (WWF), for instance, supports Community Monitoring, Reporting and Verification (CMRV) activities in Guyana. The WWF describes CMRV as a program that involves local people in Monitoring, Reporting and Verification (MRV). The conservation nongovernmental organization claims that CMRV is able to provide a cost-effective means of contributing to REDD+. The WWF supports the use of technologies, such as mobile phone applications on open-source platforms to monitor forests, for collecting information such as the size and species types in community forests.[1] CMRV activities in the North Rupununi are also supported by the Norwegian Agency for Development

Figure 8. View from the top of a hill in Surama, one of the NRDDB communities (Collins, 2014).

Cooperation (Norad), Global Canopy Programme (GCP), and the Guiana Shield Facility (GSF). Through CMRV, the communities are able to measure and take stock of their resources through the Global Positioning System (GPS) satellites. According to village leaders, the communities comprising the NRDDB have been empowered through the support of organizations like Norad. They further report an improved ability to collect and analyze data about their resources themselves.

Benefits notwithstanding, processes of this kind support the transformation of forests from historical places of refuge into commodities for conservation. After all, these forests had for centuries provided refuge to oppressed groups fleeing capital-accumulating activities taking place on the coast. Building on the earlier sovereign (de)colonial claims to the forests described in chapter 2, the commoditization and marketization of forests and the carbon stored within them further reduce the impenetrability of the forests. In its place, these processes inculcate a deeper reliance on the market for adjudicating between environmental concerns in Guyanese and Surinamese society.

• • • • •

This chapter continues in the project of tracing the colonial governing strategies that have shaped the internal affairs of Guyana and Suriname over time. In so doing, it imagines ways of uprooting them. In this chapter, I focus on how each country embraced the principles of marketization and eventually extended them to forest governance. Building on the arguments of chapter 2, which traced the emergence and shift in sovereign power that took place as Guyana and Suriname achieved independence, this chapter examines the pursuance of development models in Guyana and Suriname in the post-independence period and contrasts the strong embrace of market principles in Guyana with the reclusion and exclusion from the international community of Suriname.

As a consequence, I take a historical approach to examining the processes through which forest carbon as a commodity is created and facilitated in different aspects of Guyanese and Surinamese society. I highlight the different knowledge systems, actions, and processes through which policymakers seek to transform forests into commodities for conservation. The first section of the chapter traces attempts to commoditize forest carbon through technical processes specifically in the areas being conserved, while the second section examines how this logic is filtered throughout the wider society. All in all, I argue that the development models rooted in linear ideas of progress that position Northern countries as sitting at the pinnacle of development,[2] with little consideration of modernity's colonial roots,[3] set the tone for market-based conservation. The pursuit of these development models by successive governments in Guyana and Suriname (though more so in Guyana) instilled in wider society a reliance on the market for adjudicating social concerns.

For decades, the Guyanese government resisted externally driven imperatives that the state government submit to the strictures of the international market in the 1970s and 1980s under the leadership of the Afro-Guyanese-dominated government of the People's National Congress (PNC). In the early 1990s and 2000s, however, late in the tenure of the PNC and throughout the subsequent tenure of the People's Progressive Party (PPP), Guyana's Indo-Guyanese-dominated government, these strictures were embraced. In Suriname, while economic development ideas dependent on the market gained ascendance gradually over time, the independent state remained distant from the international community immediately after independence and engaged primarily instead with its former Dutch colonial master, from which large amounts of its development funding continued to flow.

Nevertheless, the eventual pursuit of development based on principles of modernization and marketization came to set the tone for market-based conservation through REDD+ in both countries. Through this pursuit by successive governments of both countries, a reliance on the market, which was already internally structured by racialized labor and environment relations, was instilled as the central organizing concept around which development was envisioned and made possible. Therefore, any attempt to decolonize forest governance in Guyana and Suriname requires that the market, which emerged in these two countries through colonialism as a racialized means of ordering society,[4] be shifted from its position of centrality in adjudicating in matters of environmental concern.

FOREST MONITORING

REDD+ depends on the use of technological solutions for monitoring the forest and for making patterns of its use visible and legible to those outside it.[5] Through the deployment of technological approaches to governing nature, governments and other actors supporting the implementation of REDD+ are able to take what anthropologist and political scientist James Scott calls a "high modernist" view of the forests and of the aspects of its use and management they identify as problematic.[6] In taking such a position, actors responsible for developing the system, usually based on the coast or even further afield, remain physically far removed from the subject of their governance (in this case the forests) while working towards managing it as they see fit. In so doing, they obscure the multiplicity of values associated with the forests by those within it, to install instead a logic of order imposed from above. By taking a high modernist view through technology and governance at a distance, those tasked with implementing REDD+ ignore the multiple, different use values of the forests and its users. They emphasize instead the effectiveness and visibility of the value they deem most important. Viewed in this way, REDD+ is a representation of high modernist thinking. Its adherence to this logic is

Figure 9. Annai, North Rupununi, Guyana (Collins, 2014).

demonstrated by its effort to deemphasize preexisting social and historical ways of managing and living in the forests while installing and elevating the singular use value of carbon sequestration, to be conserved through markets.

The process of making the forests legible is intricately bound up with considerations of visibility. This is because legibility, the condition of being readable and interpretable, is only possible once the object of study or management has been made visible. Visibility is, however, in the case of high modernist approaches like that of MRV, often asymmetric and nonreciprocal.[7] Much like the condition of legibility, visibility is characterized by a particular unequal positioning of the seer and the one being seen.

Space theorist Andrea Brighenti suggests that "when a transformation in reciprocal visibilities occurs, i.e., when something becomes more visible or less visible than before, we should ask ourselves who is acting on and reacting to the properties of the field, and which specific relationships are being shaped."[8] After all, "these questions are never simply a technical matter: they are inherently practical and political."[9] The increased visibility and legibility of the forests and of the effects of the activities taking place within them are similarly practical and political. This is because the illegibility that allowed the forests of the Guiana Shield to remain a place of refuge for generations is effectively being reduced. Vision is imbued with power in that everything that can be seen can be acted upon. Therefore, asymmetries in visibility are asymmetries in power.

That which is seen, is seen in ways that are in fact "socially and interactionally crafted."[10] Hence, visibility is a means of both empowerment and disempowerment. While the search for visibility by certain groups is often a quest for social recognition (for example, for repressed, marginalized, and invisibilized communities), visibility is also often "the flip side of discipline and control."[11] The potential of visibility to facilitate discipline and control is evident in the case of monitoring through technological means, since the unreciprocal nature of this monitoring results in a certain dehumanization of the observed, and possibly too of the observer.[12]

Due to its intent to pay for performance towards avoiding deforestation, REDD+ relies on MRV systems for its implementation. The word *monitor* means to "observe a situation for any changes that may occur over time,"[13] and therefore it is often necessary that a proactive role be taken towards ensuring that activities progress according to plan and that objectives and targets are being met.[14] Reporting, in the specific context of REDD+ and the UN Framework Convention on Climate Change (UNFCCC) systems, embodies the process of formally conveying assessment results to the UNFCCC according to their established standards and guidelines in order to foster "the principles of transparency, consistency, comparability, completeness and accuracy."[15] Verification, on the other hand, seeks to "ensure the validity of the information that is presented" and to ensure that certain requirements or processes are followed after a specific activity to ensure its reliability.[16] In some cases, MRV systems are used for more than just reporting on emissions reductions and removals, if other variables, such as social and environmental performance, are included within their purview.[17]

MRV enables state-making practices, operating on what could be seen as a margin of a state's existence.[18] When one considers that the state is an abstract entity made coherent through numerous recurrent social processes,[19] some of which are detailed in chapter 2, these practices of forest monitoring take on new meaning. In this sense, through MRV systems operating in support of REDD+, the margins of the strongly coast-centered states of Guyana and Suriname are further expanded to reach into territories they had not previously thoroughly penetrated. These territories, one should recall, had previously provided refuge from capital-accumulating coastal activities. Therefore, with this effort at making legible and deepening the penetration of the state comes capital.

Making Legible

Through technology, MRV systems have been developed and applied to the task of monitoring the forests in Guyana. At the time my field research was conducted, this technology was still being explored in Suriname. In Guyana, it was being presented by forestry officials working on its development and by the Norwegian government as one of the most important contributions to Guyana of REDD+, specifically the Guyana-Norway REDD+ agreement.

MRV is a means of monitoring the forest resources by drawing on satellite imagery and increasingly complicated and costly databases and equipment that should be sensitive enough to detect changes in forest cover around the globe. MRV systems are responsible for providing the requisite data for the quantification of emissions reductions and removals relative to the reference emission levels (RELs), or reference levels (RLs), that determine the amount of carbon the REDD+ participating country should conserve. The MRV system, as the name implies, involves a process for reporting and verifying the emissions reductions or removals.

MRV is carried out at multiple levels. It is, by and large, an effort coordinated by government and forestry officials in the city. It is also supported by communities themselves, as demonstrated by the earlier-introduced community-based MRV (CMRV) where local communities such as those comprising Annai are envisioned to be active, or are active, in the monitoring of forests and their use. In this framing, local communities become eyes on the ground that can help in "monitoring carbon stock changes, illegal extraction rates, production of timber and non-timber products and other variables such as biodiversity and social impacts."[20] The involvement of local communities, however, requires that these communities have their capacities built through externally led training sessions and the establishment of local monitoring systems. Local communities are also encouraged to change the way they view forests. Through externally led initiatives, the communities themselves are reimagined as actively participating in the processes of monitoring and verification. While these local communities have been functioning as eyes on the ground in different ways for decades and even centuries, their interests after independence lay in drawing attention to the forest concessions being issued by independent state governments in the forests within which they reside, rather than as verifiers of sorts, aiding the monitoring effort.

MRV systems do not address the communities' previous challenges to the state claim on the forests. In this discourse, forest dwellers are reimagined as allies in government-led forest protection efforts, instead of as challengers to their exploitative practices. Through this monitoring effort, being ardently pursued in Guyana and at its beginning stages in Suriname at the time of my study, vision and legibility is made possible both from above and on the ground.[21] These interventions shape how state governments and the indigenous communities see and interact with the forests.

These visibility- and legibility-building activities have other associated effects, however.[22] For example, some organizations involved in the monitoring effort claim that the implementation of MRV in that system could also be used for land management and the monitoring of land concessions for the extractive sectors. According to Norad, the Guyana Forestry Commission (GFC) is using its MRV system to check for compliance of concessionaires with their timber-harvesting plans, and Norad has further hopes of using it to foster compliance with several international efforts to trace forest resources. The Guyana Geology and Mines Commission (GGMC) is also using the MRV system to identify illegal mining activity.[23] Hence, the effort to monitor carbon has amounted to new ways of seeing the forests and their users and of integrating the residents of forested communities into this effort. The means through which forests are made legible in support of REDD+ in Guyana and Suriname are discussed in the following sections.

GUYANA

The process of making forests more legible commenced with zoning efforts undertaken by Guyana's British colonial master. These efforts commenced with the establishment of the Forestry Department in 1925, which eventually became the Guyana Forestry Commission in 1979. The

Amerindian People's Association (APA), a representative body of Amerindians in Guyana who lived in and with the illegible forests for centuries, tied these zoning efforts to colonialism by explaining:

> British land and development policy was primarily geared towards colonization of the interior and increased economic development, including mining development and plans for commercial farming and market gardening. Definition of Amerindian Districts was seen as part of wider land use planning needed to include Amerindians in national administration and a process for national development. As already noted, the Districts did not possess titles and indigenous peoples did not enjoy security of tenure. The British also had powers to reduce Districts without consultation and agreement, and in 1959 they dereserved 0.4 million ha of the Upper Mazaruni District to create a Mining District for the extraction of diamonds and gold (this followed earlier large-scale dereservation of extensive tracts of land in the Mazaruni Indian District in the lower and middle Mazaruni in 1933—mainly for mining).[24]

In the late 1990s, the continued existence of Guyana's forests was described by the Department for International Development (UK DfID) report as a result of neglect, rather than astute forest management practices.[25] Since then, the independent governments of Guyana have further reduced the illegibility and refuge-providing potential of the forests by increasing the use of the forests for economic earnings. The Guyana National Land Use Plan (NLUP), the state government's roadmap for managing the forest and land resources of the country, described how international thinking on forests no longer solely features considerations of logging and timber production, but now recognizes the value of ecosystem services provided by forests. These recognitions spurred the GFC to take steps towards the sustainable management of its forest resources, which feature timber harvesting guidelines, a Code of Practice for the management of forest concessions, and several other requirements and stipulations aimed at fostering good forestry practice.[26]

In 1992, the Earth Summit made visible an increased appreciation of the role of forests in the international arena, which stimulated the revision of goals, methods, and instruments to manage Guyana's forestry sector. The National Forest Policy Statement (NFPS) was prepared in 1997 followed by the Draft National Forest Plan in 2001. The NFPS described the increasing need for forest monitoring in support of sustainable forest management that required increased numbers of forest stations, mobile monitoring units, and staff. The development of MRVs as part of REDD+ activities in Guyana is in keeping with the monitoring thrust of the GFC, which is tasked with providing a performance measurement framework for the REDD+ financing mechanism.

In Guyana, MRV systems function as the basis for the identification of the drivers of deforestation. As one policymaker at the GFC explained, it was the MRV system that identified that gold mining accounts for some 93 percent of deforestation in the country, an attribution of responsibility that turned attention to small-scale miners as a problem to be remedied. Despite the problematization of gold mining in this way, gold mining has been allowed to continue practically unfettered, although some government representatives claimed that better management of this activity would eventually take place.[27]

The GFC in Guyana began the process of establishing "the world's first national scale REDD+ MRV system" in 2010,[28] which was intended to provide the basis for reporting on the country's efforts to report on the removal of carbon emissions through the use of its forests.[29] While it was being developed into a full forest carbon accounting system,[30] interim measure and performance indicators filled this data gap. Guyana, through financial support from the Guyana-Norway REDD+ agreement, was able to pay for more advanced RapidEye imagery that actually increased earlier estimates of Guyana's forest cover in comparison to measurements undertaken by different means.[31]

The development of RapidEye imagery has not been without challenges, which include difficulty penetrating cloud cover in satellite data, technical capacity building, and operational costs.[32] As a representative of the GFC explained, the agency was suffering from the lack of technical skills to carry out activities like the consultations and to develop a REDD+ strategy.[33] She depicted this as a recurring problem to be remedied by international consultants who would then build the capacity of the GFC. She explained that "to carry out reference level related activities, modelling skills are needed and there isn't the capacity to do that in-house."[34] The solution for this challenge was presented by the GSF and several other organizations that provided funding and support for building Guyana's capacity of REDD+, working towards making MRV a reality.

In sum, the demarcation of forests as distinct territories by the colonial government resulted in the increased legibility of the forests to a limited extent. During this period, the forest use practices of the communities residing in the forests had remained perceptible only on the ground, through site visits, and eventually through overhead flights for monitoring. The MRV system of REDD+ represents the newest step in forest monitoring, and while it remains somewhat limited in terms of the visibility and legibility of activities taking place on the ground in the forests, it makes the forests visible from a distance through technological means that are moving them from the realm of illegibility to legibility—from a place of refuge to a measurable and marketized site of simultaneous extraction and conservation.

SURINAME

Suriname's system of monitoring forests developed sporadically and was interrupted several times by internal conflict. Suriname's forest-monitoring organization, the Suriname Forest Service (LBB) was set up in 1947 by the colonial government to manage timber production through state concessions to lumber companies and the granting of cutting rights called HKV (Houtkapvergunning) to maroon communities. The LBB was responsible for carrying out forest inventories and collecting fees associated with timber concessions. However, after the seizure of military power in 1980 by Desi Bouterse, the country lost several relationships with foreign countries that dramatically reduced the flow of foreign investment into Suriname. Starting in 1986, armed internal conflict described in chapter 2 destabilized maroon communities. By 1986, the management of forest concessions had halted. Gold production took its place, growing dramatically in the early 1990s as the independent Surinamese state government sought to recuperate from civil war, sparked around Ronnie Brunswijk. A peace agreement was eventually signed between the warring parties in 1989, but by 1993 Suriname was economically devastated. In order to facilitate economic growth, the Surinamese government began reissuing concessions for logging with some concessions drawing international condemnation.

In 1996, the management of Suriname's forest was restructured with the establishment of the Foundation for Forest Management and Production Control (SBB), which eventually began to oversee all forest management. Forest monitoring in Suriname is now scheduled to be upgraded through REDD+ efforts with plans on stream that would increase the visibility of the forests through technologies operating in support of MRV. This improvement in monitoring capacity would increase the ability of governing actors based outside the forests to monitor economic activities taking place within them.

Imagery and data are central to forest cover monitoring, but due to a presumed lack of funding, the government of Suriname is using free data sources. According to the National Plan for Forest Cover Monitoring (FCMU) prepared by the SBB in 2014,[35] increased resolution of data is desirable, and commercially available sensors will be necessary. Therefore, funding will have to be sourced to make this possible. Additionally, improved technological capacity will be necessary to make storage and management of forest cover data possible and easily accessible.[36] Community-based MRV is also being developed in Suriname. Training in technologies related to measuring carbon stocks has been carried out with indigenous communities by SarVision to teach them to use remote-sensing technology for land and vegetation cover monitoring. According to Suriname's readiness preparation proposal (R-PP), a need exists for increased training capacity in field measurements, remote sensing, data analysis, and reporting within government organizations and in forest-dependent communities.

• • • •

In essence, MRV systems in Guyana and Suriname allow the state governments of both countries to improve the monitoring of their forests in a manner compatible with REDD+. Previously, forest management practices were based on more isolated methods in which state forest management organizations based primarily in capital cities with a few outposts in the forests carried out their work through site visits to monitor lumber or mining concessions in the forests. The implementation of MRV, however, allows for governments, along with independent, nongovernmental, and foreign bodies to view and measure the conservation of the forests, while interrogating the events taking place inside them that lead to deforestation. Consequently, REDD+ and its constituent MRV effectively enable the state to increase its unreciprocal visibility of the forests, while extending its reach or margins.[37] Meanwhile, REDD+ supporting technologies reduce the capacity of formerly illegible forests to provide refuge, by increasing the capacity of governments and other actors to monitor them.

Demarcating

In support of REDD+, aspects of the forests, now made visible and legible, are demarcated according to measurable function, such as carbon storage and land value, and then assigned a financial value. Previous efforts to manage the forests of Guyana and Suriname were centered on economic activities that led to the establishment of government bodies dedicated to forestry. Similarly, the prospect of REDD+ bringing large sums of funding to the coffers of state governments and other forest owners motivated governments in Guyana and Suriname to reorganize the state's systems for managing forests. This time, however, they were to be managed according to their carbon sequestration potential, while being measured and justified by the reliance on the technology and economic valuations of the forests.

One interviewee who worked closely with the Guyana government on the government's environmental strategy described the careful balance between seemingly competing economic activities that must be achieved in the aim of earning larger incomes:

> So in the case of Guyana and many other countries like Guyana where forestry and mining are still going to be key economic activities, we as a country, we are not going to shut down our mining industry, but what we will do is try to improve their performance and at the same time, increase their productivity, so if we could apply better practices, better technology to increase productivity, but at the same time reduce the environmental footprint, what that does is that it ensures that traditional activities like forestry and mining can continue, but it also creates that space where we can also earn from REDD+.[38]

Mining as a key economic activity, as depicted in the quote, is not to be expressly reduced by REDD+ activity but better managed through technology and improved efficiency. Through these efforts, the forests of Guyana and Suriname are being made more legible and accessible to those who seek to gain access to financing generated by the newly created conservation commodities. In so doing, they reimagine these forests in particular ways. The likely outcome of this approach is that the forests will be shaped according to the activity that would bring about the highest economic return, as explained in Guyana's evaluation of its potential for REDD+ payments.[39] However, the subsistence activities of indigenous and other forested communities are unlikely to be prioritized in this approach, because their use values do not align with the financial ones being inculcated. The entrenched interests in forest management, such as gold mining and forestry concessions, would also continue to be allowed due to their contribution to the economic earning potential of the country. Ironically, despite the identification of gold mining as the largest source of deforestation by MRV systems, this new vision of managing forests according to economic rationalities sees gold mining almost untouched, since this validation of forest use practices according to earning potential elevates it as the most important source of earnings. This consideration was reflected in the targeting of indigenous people as the source of deforestation by those responsible for managing gold mining in Guyana.[40] The issue is further complicated by the fact that REDD+ has thus far provided little financial incentive to disincentivize mining, so this economically rational model of forest management has even less chance of challenging the threat of gold mining.

Even further, a burgeoning cadre of professionals is legitimized to operationalize the technological and economic narratives of MRV, which is a process that itself facilitates the flow of capital. As previously noted, MRV technologies carry a hefty price tag, one that is usually out of reach of the REDD+ participating country. As a result, the expected costs of these activities outlined by the state governments of Guyana and Suriname are being funded largely by donor governments and international organizations. The burgeoning cadre of REDD+ professionals then draws on discussions of capacity building to legitimize their large investments into monitoring activities.

Within Suriname, the need to build capacity for field measurements, remote sensing, data analysis, and reporting necessary for effective

REDD+ implementation also holds sway.[41] In Suriname's R-PP, it is explained that the national forest monitoring system is necessary to facilitate the monitoring of all REDD+ activities by "build[ing] on available terrestrial inventory and remote sensing data, while aiming to incorporate new emerging technologies to continuously improve the quality and cost-efficiency of the national MRV system. The monitoring system will help to ensure that forests are utilized efficiently."[42] The need for MRVs is connected to the internationally established need for actions that mitigate climate change to be accountable and transparent.[43] This should, in fact, facilitate the "availability and exchange of data and experience."[44] Without the adequate operational system to fill the role of the MRV, the building of such capacity is necessitated under REDD+ preparation.[45]

The need for technical skills locally was reiterated by one official of the GFC who described that in many cases, the technical skills necessary for REDD+ were absent from the country and as a result, international consultants had to be hired. She stated: "These consultants when they come in, our staff is being, their capacity is being built as well. . . . We don't have that kind of advanced capacity to do some of the modelling and so. I would say that is one of the major constraints that we encountered."[46] As another respondent described, the implementation of REDD+ is a positive development, and so all that is missing is the mechanisms and tools to address the challenges faced by developing countries. These tools usually include capacity-building knowledge and financial resources.[47]

The capacity-building discussion and the identification of the areas of exchange of resources and knowledge between wealthier countries and poor ones act together as a microcosm of the wider development debate. It redirects attention away from the root causes of inequality, some of them colonial, towards technical solutions that come mired in increased financial dependency. The local drive for participating in the REDD+ mechanism necessitates this ever-evolving pursuit of capacity building, technological know-how, and financial resources to enable participation, while making room for the continued involvement of development organizations in these countries.

Primarily, the need for capacity building stems from the recognition of the inability of poorer states to engage in an earlier-identified activity, in this case, REDD+. Secondly, it follows from the recognition that this activity would be made possible using the technologies, methods, and tools available in more advantaged countries and that these capacities are not present in the poorer ones. An adept reviewer of this scenario would perhaps ask why it is that these countries are at such different (in the words of the mainstream development discourse, which are very fitting here) "levels of development." However, instead of engaging in that messier, historical Pandora's box that questions access to resources, exploitative relations, and so on,[48] the solution is provided that these poorer countries should simply be able to access the capacities of their richer counterparts through information and technological exchange. This is precisely the logic that fuels the continued pursuit of development in Guyana and Suriname, according to a standard and image set by more powerful countries,[49] and that continues to legitimize future development interventions as technical "solutions" continue to have damaging effects. In this case, however, capacity building adopts the technological approach to seeing the forests according to their usefulness in fulfilling a particular aim: forests as a means of storing carbon. The spread of capital is facilitated in this situation by the large international organizational architecture supporting REDD+. These organizations, together with the money and influence with which they are imbued, facilitate and are embodied by the movement of capital while furthering the aims of neoliberal conservation, ordering societies and making forests more visible and legible.

Economic Valuation

Once made visible and legible through technology and stripped of their social and historical significance, forests must then be assessed according to the economic value of their functioning in order to determine fair compensation for their services. This concern for compensation is a clear priority in the policy documents of both countries that relate to REDD+.

Guyana's Low Carbon Development Strategy (LCDS) had roots in a policy paper in 2008 that attributed an economic value to Guyana's forests based on estimates provided by McKinsey & Company,[50] a global firm of more than ten thousand consultants around the globe. The company was criticized by Greenpeace for the provision of cost curves around the world that grossly inflated rates of deforestation to give countries increased potential for continuing forest-degrading activities while gaining money from REDD+.[51] The firm is widely used in climate change and environmental matters to provide economic analyses and to attribute financial values to environmental services. This value estimation acted as the foundation of Guyana's REDD+ effort, which builds on McKinsey's estimates to estimate Guyana's forests' value to the world at US$40 billion per year.

The Guyana Government explained:

> Our work suggests that baseline assumptions should be driven by analysis that assumes rational behaviour by countries seeking to maximize economic opportunities for their citizens (an "economically rational" rate of deforestation). Such baselines can be developed using economic models of expected profits from activities that motivate deforestation (vs. in-country benefits of maintaining the standing forest), and timing and costs required to harvest and convert lands to alternative uses.[52]

The estimated value of those in-country benefits was estimated at a more conservative US$580 million per year. The economically rational path that Guyana should take was depicted by a wide array of statistical graphs based on economic valuations attached to activities that have traditionally taken place in the country, or that are likely to take place to generate income. For example, the estimation of the carbon abatement costs for predicted avoided deforestation in Guyana amounted to an annual payment of US$430 million to Guyana for the services of its forests. It is worth pointing out that these values are estimates based not on historical trends, but on possible future pressure on the forests. Development here is used to justify the need for these policy shifts, and for REDD+, since a "rational" development path is predicted as necessitating the destruction of forests. Therefore, the pursuance of REDD+ through the LCDS draws on these economic rationalities rooted in neoliberal logic and points out that a rational development path would result in the destruction of the forests, making room for REDD+ to alter that equation. Thus far, only forests conserved and managed by the state have been allocated for REDD+ activities. Indigenous groups who have some tenure over the forests within which they reside (in the case of Guyana but not in Suriname) should eventually have the option of opting into the REDD+ mechanism and being remunerated for the services of their forests.

In Suriname, due perhaps to its later move towards REDD+, the reliance on the economic narratives was less overt. Their process of REDD+ preparation took a different path commencing with their R-PP. Given that Suriname was operating within a preestablished framework that outlined the funds that would be available for REDD+ readiness, its R-PP was prepared in response to this possibility. Suriname did not go the route of having its forests valued and offered up for incentivization by the international community. There were no bilateral agreements with Suriname to generate funding for MRV system development. Suriname was very much in the preparatory stage of REDD+ when the data for my project was gathered, accounting perhaps for why a reliance on technology and economic valuation was less present in their R-PP. Instead, Suriname presented a listing of the potential future drivers of deforestation in its R-PP to contextualize the incentive potential of REDD+ in the country, including mining, logging, infrastructure development, agriculture, energy production, and housing development.[53] The reliance on associating economic values with the work of the forests is not limited to MRV and forest management practices, however. In the following sections, I move towards demonstrating how this market logic is expanded throughout the wider Guyanese and Surinamese societies.

CENTERING MARKETS

Money Does Grow on Trees—The Environment Reimagined as Natural Capital

As I observed elsewhere, indigenous and maroon communities of Guyana and Suriname are increasingly seeing the forests no longer as primarily a provider of sustenance, but through the transformative lens of the market.[54] Made possible through the ever-present promise of development, this change is made evident through both the restructuring of their social relations with their environments, and their internal interpretations of themselves. Within discussions on REDD+, the different means of gaining income while conserving the environment often feature suggested replacements to deforestation, such as ecotourism, sustainable forestry, and the sale of nontimber forest products. Hence, forests and the environment in both countries are being reenvisioned in these communities as natural capital. This is demonstrated by the way persons on the ground refer to the worth and value of the forests. In order to prepare communities for REDD+ implementation, overt efforts are carried out by the governments of Guyana and Suriname to facilitate their absorption into the dominant economic frame of the society. In other words, in keeping with the logic of the neoliberal subject of *homo economicus,* the preconditions for his or her functioning are being set up in these communities.

The process of establishing the preconditions for a functioning market in indigenous communities had commenced long before the introduction of REDD+. According to Griffiths and Anselmo,[55] most Amerindians in Guyana are earning some amount of cash and survive on a mixture of cash and subsistence activities though levels of dependence vary between communities.[56] Actors related to REDD+ in Guyana frequently refer to the need for the introduction of the cash society in the rural areas,[57] referring to the removal of the use of barter and the installation of economic relations based on the exchange of cash in its place. REDD+ is explicitly described by policymakers and community members as a means of bringing cash to indigenous communities, a welcome proposition to the communities themselves.

However, to qualify for this cash, and to be able to manage it, certain changes must be made to the way the communities are being ordered. The introduction of the cash society is seen now by some indigenous leaders as an incentive for the destruction of forest, since forest, a resource whose value was previously difficult to exchange, is now represented financially. Joe, a representative of a group of indigenous community leaders that is successfully implementing REDD+ forest conservation activities, explained:

> Indigenous people are custodians of the forests. First, they used to barter up to approximately 15 years ago. Now, there is a cash society and if there is no payment for the ecosystem services, people, even the indigenous people, will destroy it because they are becoming greedy also.[58]

As one representative of this indigenous group in Guyana pointed out, Annai has huge tourism potential with different tourist attractions available in the different constituent villages, including cultural and landscape tourism, birding, and activity-based tourism.[59] His comments reflected the emergence of a situation where indigenous people started to see themselves as outsiders looking in and were determined that their culture would be a spectacle that would be able to earn them an income.

Both within and without indigenous communities, market considerations were a pivotal factor in discussions on REDD+, forming the central language within which policymakers and other decision-makers couched their approval or disapproval for the mechanism. The introduction of the cash society seems to indeed be contributing to shifting indigenous views of the forests from environmentally oriented to economically dominated. The most striking example I encountered was that of a REDD+ crew member in the North Rupununi who, when asked what he thought REDD+ was for, replied by rubbing his thumb and index and middle fingers together to demonstrate that it was about money.[60] The fact that this community member described to me just the financial component could indicate this changing view of the forests from one of sustenance provider to one of capital, a shift that begs the question of how their relationship with the forests will change when they continue to see them as natural capital without sufficient earnings from REDD+.

Policymakers also tended to view nature as capital and supported the financial valuation of nature as a basic step for making land planning decisions. A former commissioner of the GGMC highlighted this centrality by explaining his assertion that land use planning and environmental management should be dependent on the unit value of land. He explained that the revenue brought in by tourism, rice, standing forests, and mining, for example, should be known before decisions are made on which activity should be pursued.[61]

According to representatives of the group of indigenous community leaders, in order to facilitate the growth of the tourism industry in the area, there is a need for cost-effective and efficient development of infrastructure to reduce the capital outlay.[62] This desire to gain as much income as possible by having hundreds of guests per year is being tempered by concerns for the environment. There is clearly an awareness even in this reframed view of the environment that "the thing about this tourism thing is not for you to get rich, it's for you to sustain yourself, in the meantime you build your community and support community efforts like schools, clinics, whatever, scholarships."[63] It was clear that even while communities in the Annai area are increasingly being integrated into the capitalist economic system and relating to the environment as primarily a source of income, their awareness of the need to conserve remains somewhat in place.

Indigenous communities in Suriname face similar options. In Apoera, Suriname, they too are flirting with the option of gaining income from ecotourism, and they often compared their natural endowments with those of more widely known tourist destinations in Latin America.[64] However, they are more constrained in their effort to transform their forests into sources of income due to the absence of formal rights to the land, since the rights to the land in the post-independence period are vested almost exclusively in the Surinamese state, a situation detailed in chapter 2. Therefore, outside of the practice of engaging in extractive economic activities themselves, indigenous groups and maroon communities in Suriname are less complicit in the effort to envision and depict the forests as natural capital. Given that private rights to the land represent one of the first steps in setting up a market-centered economy, I surmise that the neoliberal concept of natural capital is less readily accepted in Surinamese forest communities.

At the policy level, REDD+ is being proposed as a way of bringing increased revenue to the countries by raising the country's profile in terms of ecotourism.[65] The transformation of the environment to "natural capital" has permeated all levels of society, with conservation organizations leading the charge. The head of one of the organizations explained: "So, we help countries first and foremost, identify natural capital and determine how to use it successfully to leverage their development . . . and then we help countries, once they have decided that they want to protect an area or manage it carefully."[66] This representation of the environment as natural capital filters the market-based method of managing nature throughout different levels of society and lends a technocratic feel to discussions on the environment.[67]

Capital is widely associated with markets, investments, and economies, a realm dominated by rational thinking, logic, and figures. The forests of Guyana and Suriname, which had been, in colonial times, largely incidental considerations in the dominant interventionist and capitalist narrative, have since been reshaped in the imaginations of city residents and the international community as a store of raw materials, and as capital that can be managed, accounted for, utilized, invested, reaped, and represented numerically. Earlier sovereign and political relations with the forests are now being tempered by the rationality of economics and neutral image of markets in a way that disguises the fact that global markets have, in fact, generated the threat to the forests through demands for minerals and timber resources.[68]

The view of nature as natural capital is touted by development and international organizations such as the World Bank and the United Nations Development Programme, but notable here is the fact that even at the community level it has changed how residents see the forests and their relation to it. This constant reshaping of the image and discourses around the forests is largely the effect of the international flavor of the day, one that now targets the remedying of environmental concerns while achieving continued economic growth. At times, the stated environmental concern is overtly economic, as illustrated by the assertion by the previously mentioned conservation NGO director's view that one sustainable approach for Suriname is that of selling bulk water to foreign countries since the forests are not commercially valuable enough to bring in enough income.[69] Conservation-oriented organizations are themselves engaging in policy advocacy and taking positions on how to manage an economy and to earn income and stimulate economic development, with conservation itself taking a secondary role. This view of nature as a means of earning income is entrenched also in Guyana, as one policymaker within the GFC insisting that REDD+ and the payments it generates are "not a loan or grant, it's a payment for a service which we are providing."[70]

In justifying REDD+, the policymaker further explained that forests cannot be conserved without payment, since without payment there will be utilization, and this utilization could be unsustainable depending on the pressure of the day.[71] The implication here is that without financial incentives for conservation, forests are likely to be destroyed. This logic ignores the current power of sovereign rules and regulations to manage the forests through national public funds, and it lumps all groups associated with managing the forests as responsive to market pressures. This sidelining of governmental authority in the management of natural resources was evident also in the suggestion of a representative of a civil society organization representing business interests who, while pointing out that public organizations such as the aforementioned GFC are charged with managing and protecting forests of Guyana, suggested that REDD+ was not that different, since it is simply the international community paying for the management of the forests from which they all benefit. This imagined extension of the forests as a global good brings with it ideas of global responsibility and shrinks the role of the state governments of the country within which REDD+ activities are taking place. Meanwhile, this extension sidelines the relations, both historical and current, that local communities have with the forests. The market expands its reach to bring public goods under its remit,[72] while state control and funding of the management processes within its borders shrinks, making space for the continued spread of the market as the primary mode of distribution.

Selling Nature over Lunch

At the offices of an international conservation organization working on activities related to REDD+ in Guyana and Suriname, I participated in a few lunch table conversations that provided insight into their organizational priorities. It was clear that the staff of the organization and I shared concern for the interaction of human beings and nature. We believed in simultaneous use and nature conservation and hoped to find a balance between competing uses to satisfy human wants/needs and conservation. The staff members of the organization insisted, however, that we engage with the major polluters in the world to make a difference. The director of the Guyana office explained that we should seek power out and engage with it within the current hierarchy. He stressed engagement and dialogue with world leaders and responsible businesses, in the firm belief that improvements can be made in humans' management of the environment through these efforts. He further asserted that when indigenous populations live in pristine environments yet cry out for educational opportunities and means of "development," the resources they desire must come from some other natural environments somewhere on Earth. Since all human "wealth" and progress comes directly or indirectly from the Earth, some sacrifices must be made. The challenge, according to this view, is in finding a tolerable balance between natural resource use and conservation.

This organization supports REDD+ activities in both Guyana and Suriname. In Suriname, it is spearheading its own contribution to the REDD+ implementation process, which is intended to build the capacity of the government of Suriname to engage with stakeholders, especially indigenous and maroon communities. However, this organizational effort has been fraught with controversy, since some indigenous groups claim to have come up with the idea themselves only to have it refined and re-presented to them by the conservation organization without acknowledgment of their input.[73] The project's progress had also been challenged by an indigenous representative organization that sees the conservation organization as an untrustworthy partner due to its director's recent change from government representative to head of the international conservation NGO. Essentially, the indigenous representative organization sees itself as being unable to transparently engage with the conservation organization due to the number of hats worn by its director.

This organization's director in Suriname was much more explicitly market focused than his Guyanese counterpart. A gap existed between the engagement and facilitating function claimed by the NGO on its website and the economic focus and contentious relationship with indigenous representatives in Suriname. During a REDD+ conference hosted by the state government of Suriname, the organization's director, in presenting his view of REDD+, heavily emphasized the economic benefits that could be gleaned from REDD+ activities and from monetizing nature. His presentation focused on what he described as a climate finance challenge of gaining enough financial support for REDD+, a problem that he saw as rooted in the lack of demand for carbon credits. The director pointed out that investing in nature was a cost-effective solution. He made this claim by citing data provided by McKinsey Global Institute.

The director in Suriname went on to propose, as previously mentioned, bottling and selling the freshwater resources of Suriname and engaging in large-scale export of the water. He argued that this was the next frontier for economic growth in Suriname, while ignoring the environmental downsides of these activities, such as plastic pollution, carbon emissions of transportation, and the high consumption of oil in producing bottled water. In other communications with me, he expressed a very pessimistic outlook on the prospects for combating climate change on the global level and proposed that since the damage is done with little hope of carbon emissions abating in the short term, what is necessary is that Suriname better utilize its natural resources to gain funding for the strengthening of its infrastructure to deal with the inevitable impacts of climate change. This type of rationale demonstrates how some key conservation NGOs draw heavily on the narratives of economic growth as a facilitator of combating climate change. In this approach, environmental concerns are no longer the overriding factor but have been superseded by the exigencies of capital expansion, blurring, or in fact eviscerating, the line between conservation and development.

REDD+ fits neatly into that organization's mandate, with Suriname's director stating that "we see REDD+ as a very clear example of how to actually look for alternative ways that leave natural capital intact that still allows for development."[74] Looking past the inevitability of the forests' destruction implied in this statement, this transformation of nature from the focus of conservation efforts to its depiction as natural capital is reflective of the commodification of nature and the dominance of the economic growth imperative in REDD+ implementation. Economic growth is the priority and is seen as a means to environmental conservation, even though on the way, some conservation NGOs become effectively as economically oriented as that which they purport to fight.

Conservation organizations contribute actively to the marketization of nature, with an explicit focus on economic growth through nature conservation and use. This overt subscription to the view of nature as capital has been challenged in academic literature. Chapin questioned the current modus operandi of these big international conservation NGOs,[75] within which power, through finance, connections to government, and big business, is imbued.[76] These concerns of competition and covert interests remain valid, since contrary to my expectation of encountering a proliferation of new institutions to support the preparation for REDD+, I found that the mechanism had, in large part, been captured by preexisting institutions and organizations that work on development and environmental conservation in Guyana and Suriname, and around the globe.[77]

REDD+ continues in the colonially rooted vein of representing Guyana and Suriname as stores of value to be exploited. It represents forests as "fictitious commodities" that act as natural capital and stores of forest carbon for the global good and that enable the abstraction of value from their natural resources.[78] While REDD+ is purported to embody a new, more effective way of managing the globe's remaining tropical rainforests, the mechanism facilitates continued exploitative, resource-driven engagement with these countries, to which has been added the fictitious commodity of carbon.[79] Through REDD+, the forests and the communities that historically sought refuge within them, along with the wider society, are reenvisioned in ways that make them amenable to the spread of capital and are, in some cases, themselves being shaped into capital. Hence, even in the post-independence period, Guyana and Suriname continue to be imagined as a store of resources, an imagination that has moved from plantation-based economies centered on sugar and other crops, to extractive ones based on lumber, bauxite, and gold, and now to conservation-oriented ones, all while going through a process of neoliberalization,[80] a process I flesh out further in the next section.

Aligning with the Global Economy

REDD+ goes beyond its previously described contribution of changing the relationship between indigenous communities and the forests and is now depicted as an income provider, one that further integrates Guyana into the global market economy. REDD+ is portrayed by its proponents as having the ability to generate economic growth in excess of Latin American projected growth rates, while guiding the economy of Guyana specifically along a low-carbon pathway. The Guyana government, in its R-PP, describes the mechanism as oriented towards ensuring that those persons who derive economic benefit from the use of the forests can continue to do so in more sustainable ways established through international practice.[81]

Interestingly, the current orientation of development policy in Guyana has barely shifted from its neoliberal underpinnings and market focus depicted in policy documents of earlier years. The shift towards environmental concerns is almost imperceptible in the framing of the challenge, future development goals, and aims of the country. The government of Guyana, again in its R-PP, states that the overall aim of REDD+ payments will be to enhance the economy through the attraction of investment into areas such as "human capital" and social services,[82] while using payments for job provision and higher standards of living. The government explains: "Transforming Guyana's economy will require striking a balance between attracting large, long-term private investors who will have a catalytic impact on the national economy, and making significant investments in human capital and social services to equip the population for participation in the new economy."[83] The economic growth orientation of REDD+ could not be clearer, with its references to human capital, natural capital, and explicit stating of efforts to ready the population for participation in economic activity. The R-PP continues:

> It [REDD+] will also require a balance between using forest payments to enhance the opportunities for those who live in the forest and recognizing the rights of other Guyanese citizens, including the urban poor. The importance of benefit sharing with Guyana's Amerindian communities is particularly important. To meet the needs of both forest dwellers and the population at large, Guyana will invest a significant share of the forest protection funds it receives in initiatives aimed at developing jobs and diversifying the jobs base, and improving the general standards of living of its citizens.[84]

REDD+, as shaped in Guyana, is a tool for development. Considering that REDD+ is intended to shift the economy along a greener and more low-carbon path, it is telling that the rhetoric underpinning the agreement, like that quoted above, differs little from that of the National Competitiveness Strategy (NCS) published in 2006, well before the entrenchment of market methods of conserving the environment. This demonstrates the pervasiveness of neoliberal ideas and how little they have changed in the Guyana government's reinterpretation of REDD+. This emphasis on attracting investment and instilling discipline in its people was evident even then,[85] as the following quote from the NCS demonstrates:

> Important progress has been made in Guyana in recent years in managing the process of adjustment to the new world economic environment through exercise of monetary discipline, improvements in the environment for private investment, reform of the tax system, creation of a property market, investing in basic education and infrastructure, and boosting productivity in traditional sectors of the economy. But important and pressing challenges still remain to be tackled. At the forefront is achieving the economic imperative of improving national competitiveness and diversifying the economy.[86]

The principles stated in Guyana's NCS, from an era preceding REDD+, are remarkably similar to those espoused in its REDD+ preparation documents. This is demonstrated by their shared emphases on boosting productivity and fostering national competitiveness. Economic growth through foreign investment and market discipline has, in recent decades, been the go-to tool for development in Guyana, and the shift of rhetoric from economic development to sustainability is belied by the continued reliance on market tools and methods for ordering the society.

The difference of the LCDS, underpinning REDD+, is the shift in focus from the capital centers to indigenous communities previously marginalized in development discussions. The logic, however, remains the same—that economic growth through the market is central, with no shift in the economic activities deemed necessary for development. With this in mind, the spin-offs of REDD+ become clearer: indigenous visibility, legibility of the forests,[87] the reimagination of the forest, from obsolete to suitable for extraction (store of natural resources to be exploited) and now to representative of financial value through forests reimagined as natural capital.

REDD+ continues in the vein of instilling self-discipline, market awareness, and efficiency in the society as depicted by previous policy documents upon which Guyana's official development trajectory was based. REDD+, as reinterpreted by the state government of Guyana, is seen as adding to the basket of economic sources with little engagement with how these sources conflict with each other. The NCS does, however, attribute some recognition to the international drivers of resource use, stating:

> Moreover, the situation is now more critical than ever because the mainstay of Guyana's economy, Sugar, is under threat from the EU's reform of the Sugar Protocol which will reduce the landed export price received for Sugar by 36% over four years and could potentially amount to a loss equivalent to 5.1% of GDP and 5.4% of merchandise exports annually.[88]

Interestingly, the NCS points out that the need for Guyana to compete on the international market and to diversify its base is linked to the removal of preferential trade in historical stronghold commodities that, as previously pointed out, have driven the very creation of the modern-day state of Guyana. This is also true in the case of Suriname, which according to my respondents has already embarked upon a path of diversification due to the people's insistence on disassociating themselves from the production of sugar and other historical goods.[89] This diversification has resulted, however, in a reliance on remittances from the Surinamese residing in the Netherlands,[90] extractive industries, and the state as a major employer,[91] as discussed in chapter 2. However, this challenge to historically entrenched income-earning activities and identities has shaken Guyana to the core. Certain ethnic groups, in this case the former indentured servants, are particularly threatened by the ongoing reduction of their livelihood options through the recent economic unviability of sugar. As the NCS demonstrates, much of the need to be competitive and to instill neoliberal logic of efficiency and cost-effectiveness throughout society is spurred by this need for financial viability at the country level.

Additionally, the shift of former colonial centers in Europe towards deepened relations amongst themselves in the European Union (EU) has affected the viability of sugar exports. Their integration resulted in a coalition of states, which as a unified entity, is no longer definitively characterized by colonial histories and relations. The desires of a now-unified (at least from the viewpoint of Caribbean states) Europe have reduced the intensity of the discourse of colonialism and post-colonialism and have left small, now-independent former colonies without preferential access to markets, fighting a battle to adopt the characteristics they see as associated with the strength of their former colonizers. It appears that countries like Guyana see the need to adopt the weapons of the strong, while shifting their rhetoric and policy orientation to reflect the international concerns du jour, if they are to win the battle.

The need to diverge from this reliance on monoculture and dependency on sugar had been recognized by the government of Guyana by 2006 with the publication of the NCS. Ironically, the lack of incentives to change the development path was the topic of the day:

> Guyana relies too heavily for its economic existence on the production and export of a few virtually unprocessed commodities. In other words, the country's economy is almost totally dependent on the production and export of raw materials. Moreover, most of these products are sold in guaranteed preferential markets at prices which even now are generally higher than

> those that are obtainable in the non-preferential world. As a consequence, the Guyanese producer has had no *incentive,* indeed no overwhelming reason, to be competitive, to be as efficient as possible. . . . Moreover, because of the ready acceptance of certain of our export products in these favourable conditions, we have tended to concentrate on only a few products, and to continue to employ outmoded production practices.[92]

The recognition that preferential trade would soon be removed brought about an impetus towards diversifying the economy.[93] Thus far, it seems that gold production has filled that gap, with its contribution to Guyana's GDP rising from 7 percent in 2007 to 12 percent in 2010 and continuing to over 15.5 precent in 2011,[94] with detrimental effects on the global environment. The need to neoliberalize the economy could not be more explicit:

> Competitiveness, at the micro level, means the capacity of our firms to compete, to increase their profits and grow. It is based on costs and prices, but more vitally on firms' capacity to use technology and on the performance of their products determined by a wide range of factors. . . . Building national competitive advantage is not a matter of a fixed production structure predetermined by a given and unchanging set of endowments. Competitive advantage does not simply exist. It has to be *created.* It has to be carved out of initial conditions through the right enabling environment, through conscious investments in technology, education, training, information search, engineering and even research and development to create new skill and technological endowments that can allow the economy to grow by diversifying and deepening the productive base.[95]

Competitiveness must be created, according to the government of Guyana of 2016. As previously noted, the largest driver of deforestation in Guyana and Suriname is gold mining, followed by logging and smaller economic activities like infrastructure and subsistence agriculture. International markets are the main destination of the products of forest-degrading activities. In Guyana, the year 2015 saw 450,873 ounces of gold being declared by gold miners, with 448,248 ounces being exported, demonstrating that 99.41 percent of the gold declared was destined for overseas markets.[96] The logging industry is less export oriented, with 137,407 cubic meters of the 336,318 produced (40%) exported in 2015.[97] In 2014, Suriname declared the production of 30.782 kg of gold, of which 30.034 kg was exported (97.57%).[98] In terms of the production of goods, domestic markets are not by any means the major consumers of the items that contribute most to deforestation. These activities are subject to the demands of the international market. In essence, the loss of preferential markets in Europe, coupled with a lack of vision and execution towards diverting the economy away from its historically charted past, resulted in an overt and explicit adoption by the state government of Guyana of neoliberal tendencies and market dependence, which then had to be instilled in the society to develop an efficient workforce with an enabling environment.

Suriname's eventual move towards embracing market principles was demonstrated in its Country Partnership strategy with the World Bank, which was developed as the country prepared for the suspension of the Multi-Annual Development Plan in 2011 in a bid to access new development partners. The Suriname–World Bank engagement focused on "public financial management (PFM) capacity building, fiduciary improvements, competitiveness, and social development" as the government sought to improve "economic diversification to decrease dependence on extractives and increase the resilience of the economy."[99] Thus, through their quest to get ready for REDD+, the Surinamese state government sought to simultaneously address concerns for the environment and demands for economic growth.

Yet, among REDD+ stakeholders in Suriname, there was a comparatively weak imagination of the market as the adjudicator of environmental concerns. REDD+ and its progress there were rooted instead in development imaginations and the need to gain funding to achieve development. Hence, Suriname's turn towards REDD+ can then be situated within a national turn to external sources of income to facilitate development and a gradual, though resisted, move towards neoliberal development models as the post-independence payout from the Netherlands dried up or became too onerous for the government to pursue.[100]

CONCLUSION

This chapter traced the processes through which forests moved discursively from a place of refuge to a store of value that can be drawn on to support economic growth and development. In so doing, it presented a means through which decentering the market can be envisioned, that is, primarily by resisting the tendency to have the market function as the primary adjudicator of social and environmental concerns. Building on the historical state claim detailed in chapter 2 through which the forests were claimed for various extractive practices in the colonial period, the state governments of Guyana and Suriname are improving their ability to manage the forests from a distance through modern technologies. Through these technologies, they are able to make populations within the forests and the effects of the activities in which they engage increasingly legible and visible. This increasing legibility and visibility of the forests is accompanied by processes that demarcate aspects of the natural environment according to their economic valuation and those that support the pursuit of these activities according to their potential contribution to state coffers. However, the logic that enables forest carbon to be commoditized, albeit unsuccessfully,[101] is not limited to the forested areas. This logic is instead actively expanded to the wider society through internal and externally directed policy efforts and organizational priorities. The adoption of this logic is evident even in conservation-oriented organizations, which also work towards instilling a reliance on markets for adjudicating environmental affairs.

Yet, these efforts are not accepted passively, as some actors in both countries, within and without the forests, recognize the need to decenter the market. In the following chapter, I highlight the means through which different groups identified as stakeholders of REDD+ question, challenge, and resist the increasing marketization of society. I understand this resistance to embody an undisciplined response to environmental governance,[102] one that should be taken seriously and carefully navigated in what I identify as the third step towards decolonizing environmental governance in the Amazonian Guiana Shield.

4 Undiscipline the Subjects

> For monitoring purposes, *traditional knowledge of local communities needs to be converted to western knowledge.* Communities need to be trained in these specific requirements and to monitor field activities in an informed and systematic manner to deliver data that can be incorporated in monitoring systems.[1]

This chapter focuses on the disciplinary logics that support the implementation of the Reducing Emissions from Deforestation and forest Degradation (REDD+) mechanism. Disciplinary logics support governance strategies by instilling in the population of its governance certain norms and values in the aim of altering behavior. These behavior changes should ideally then support different aspects of the REDD+ initiative. As demonstrated in the quotation above from Suriname's REDD+ readiness preparation proposal, one such desirable behavior change is the transformation of traditional knowledge systems of local communities in the forests into Western knowledge systems.

Regardless of where it is being implemented, however, REDD+ relies on the provision of financial incentives for bringing about behavior changes in forest users. In so doing, it conjures up an image of a deforesting individual, company, or even government that responds well to such incentives. Given its global pursuit, the incentivization logic upon which REDD+ relies has successfully appealed to a wide cross section of stakeholders.

However, the racialized subject of colonial governance introduced in chapter 2 was disciplined into emergence across five centuries of colonial governance. I use this notion of the abstract racialized subject as a counterbalance to *homo economicus,* which is the subject known through governmentality theory to be amenable to the market-based governance invoked by REDD+.[2,3] Singular in nature, the racialized subject represents an idealized subjectivity, or personhood upon which different characterizations of race are projected. This singular subject differs markedly from its possible pluralization, that is, racialized subjects. This difference is rooted in the fact that I see pluralizing a noun as cementing, to some degree, the characteristics associated with that noun.[4] In referring only to a singular racialized subject, I minimize the essentialization of people through characteristics congealed in racial categories and hold the door open for their racialized subjectivity to be recognized as shared, fluid, historically conditioned, and unessential to their beings.[5] The racialized subject is characterized by their racialized relation to capital-generating activities in the present and remains dominant in Guyana and Suriname even after independence.

The process of forming, disciplining, and/or undisciplining a subject, racialized or otherwise, is by no means simple. This process should ideally allow the subject to successfully internalize and act upon externally introduced norms and values. The accurate identification of any one factor to which any person responds, in determining their behavior, is a notoriously difficult, if not impossible, task. In view of this limitation, I take up a search that is not causative in its pursuit of identifying where *homo economicus* exists, but that instead identifies where *homo economicus* does not exist.

I carry out this search in the group identified by both governments as REDD+ stakeholders. I examine the views and interpretations of REDD+ and forest use activities they expressed in dialogue with me and in public. I do so while paying careful attention to the multiple ways in which people who have been grouped together as stakeholders in REDD+ preparation documents, whether based on racial, organizational, or occupational markers, respond to the processes of legibility building, state claims, and marketization highlighted in previous chapters. In engaging with their views on REDD+, I highlight the incoherence of their views and responses through the use of the term *fragmented subjectivity,* a concept that points to the contradictions that exist between the ways stakeholders are represented in policy documents and their expressions and subjective identifications. Notwithstanding post-independence efforts to shape *homo economicus* into being, I find this subjectivity largely absent, ergo I find REDD+ untenable. For REDD+ proponents, however, the absence of *homo economicus* in these areas further justifies the use of disciplinary methods for cultivating this subjectivity into being, as the opening quotation of this chapter demonstrates.

This chapter therefore challenges the clarity and simplicity of REDD+'s incentivization logic by offering up instead an overview of how REDD+ stakeholders in Guyana and Suriname respond to REDD+ as a mechanism of environmental governance.[6] I find that these stakeholders do not conform to the image of a *homo economicus*-like deforesting subject who will change their behavior according to incentives provided. Instead, they span the spectrum of undiscipline. This chapter further demonstrates how the tendency to essentialize groups of people related to the REDD+ mechanism in particular ways allows for blanket suggestions that further the interests of REDD+ in the international arena. The essentialization of the subject of REDD+ governance ignores the number of factors that contribute to shaping the very identities of these actors, which amount to much more than just economic considerations. As I observed elsewhere,[7] in response to the disciplining tendencies of the market, the undisciplined subject fragments or simply refuses to be disciplined.

During the colonial period, the European colonizers of Guyana and Suriname ensured that capital would continue to be accumulated through the appropriation of natural resources in the Guiana Shield by disciplining the behavior of the diverse groups of people they encountered and brought under their control. The successful internalization by the colonized of these norms and values cemented the colonizers' hold on power. It also cemented their access to natural resources and their position at the pinnacle of these hierarchically stratified and racialized societies. This outcome was especially important to the colonizers' interest because, over time, they had become significantly outnumbered in these colonies by those they subjugated. Thus, decolonizing forest governance depends not only on the identification of the social norms and values instilled in the Guyanese and Surinamese populations that acted in support of the colonial enterprise, but also on their subsequent dismantling. In other words, the subject of decolonial governance must be undisciplined.

QUESTIONING MARKET DOMINANCE

As a process, governance depends on a series of interventions into the population that has been identified as its subject. In the Foucauldian framework I adopt throughout this book,[8] governance is not limited to its sovereign sense, for example, elected state government officials managing the affairs of independent states. Instead, governance can be undertaken by a wide variety of unconventional actors, for example, parents managing families or prison guards managing inmates. The openness of this framework contrasts with fixed interpretations of governance as being the exclusive domain of certain fixed actors. Instead, governance in the Foucauldian sense examines how different actors become imbued with the ability to govern and the strategies through which they go about doing so.

Think, for example, of a kindergarten class being managed or "governed" by a teacher. In recognizing the age, propensity for tiredness after physical activity, and dietary preferences of the class, the teacher might schedule nap times, apportion treats to reward what he identifies as good behavior, and lead learning exercises to regulate individual behavior while moving the students towards a particular outcome. In this case, that might be learning their alphabet and developing collectively harmonious behavior patterns. In recognizing the characteristics of this student population, the teacher, classed as government because he is instilled with the power to manage this particular population, devises ways to engage with and to intervene in the existing behavior patterns of the population to achieve his aims. In so doing, he draws on a series of tools to influence the population's behavior (treats as rewards), knowledge systems (scientific studies that say that children of kindergarten age get tired after 60 minutes of physical activity),[9] and judgments (socially harmonious behavior is a desirable character trait) to nudge the kindergarten population towards a particular aim.

Similarly, (de)colonial mentality as strategies for both colonization and decolonization recognizes that the colonial governments of the colonies that became independent Guyana and Suriname used an assortment of less palatable tools that drew on different types of knowledge, judgments, and ways of doing things to achieve their desired outcomes. Given that governance is not limited to human subjects, I am applying the tenets and principles of this framework to an analysis of the governance of forests and their users.

Governance, in this sense, is not without limits. Anthropologist Tania Li described its four limits,[10] noting that, first, the art of government should not be conceived of as totalizing, since people retain the possibility of acting otherwise. Drawing again on the kindergarten example, this limit recognizes that despite the teacher's best efforts, a child might not get tired after running around at break time. Hence, the balance the teacher seeks to achieve is always in relation to the agency of the children in his classroom.

Second in number but not in significance, governing interventions cannot be seen as the sole factor in determining an outcome, because life is unpredictable and social relations and histories cannot be simply erased. In our kindergarten example, this might mean that domestic abuse at a child's home might hamper his or her openness to learning or playing on a given day or year. Third, governing interventions should not regulate social interactions in a totalizing manner but should instead intervene as little as necessary to achieve the desired outcome. This is because the intervention itself is likely to stimulate certain associated, and often unpredictable, side effects. This third limit might manifest itself in the teacher opting to intervene in existing behavior patterns by relying on treats as he develops systems of reward and punishment in place of time-outs. The kindergarten teacher might make this decision after weighing the pros and cons between cavities as a possible known side effect of sweet treats, and socially isolating behavior as a possible known side effect of time-outs. He might then determine time-outs to be more harmful to his aims.

Fourth, governance is limited by the effort to transform issues that would otherwise be addressed through politics into technical areas of efficiency that are devoid of their political context. This fourth limit might see itself manifested in our teacher finding it difficult to maintain perfect levels of class attendance by providing penalties for absence (a technical intervention), because he has not considered how domestic abuse at home (a political issue outside his control) affects the child's ability to come to school. As Li described,[11] attempts to transform the realm of the political to that of the technical will never be complete.

Returning to the forests of the Amazonian Guiana Shield, resonances of the first limit of (de)colonial mentality can be seen in recent efforts to govern the forests through REDD+. These and other policy attempts to extend the logic of the market to the management of climate change and the environment have not been accepted by all factions of the Guyanese and Surinamese society. People across stakeholder categories expressed their concerns about the mechanism and its governing logics. These concerns ranged from how REDD+ is being implemented, to the adequacy, or lack thereof, of basing environmental policy decisions on monetary values. Even actors who are themselves leading the charge to implement REDD+ in these two countries questioned the mechanism's aim and modus operandi to some degree.

The United Nations Development Programme (UNDP) had a pivotal role in implementing REDD+ in Guyana and in Suriname. Yet, within its leadership, concerns about REDD+ were expressed. My respondent in the Suriname office (UNDP-S) was particularly concerned about how REDD+ was being funded. He argued that markets are not suitable for addressing the problem of climate change. He added that this was because climate change is not a problem of demand and supply, because there is no demand for carbon emissions. Ironically though, his phrasing of his objection reflects the dominance of market logic in common parlance. He explained that he is not supportive of the idea of buying and selling carbon credits. To me, this was an interesting declaration because, according to him, the majority of the organization's funding was at the time tied to REDD+. His exact phrasing was as follows:

> I am not big on the idea of buying and selling carbon credits. It might sound very strange from someone like me. I find this to be an orthodoxy that will just later down the road, keep us wallowing in the same situation that way. I have tried my best to expose myself to the literature on carbon credits. I am not sold on it.[12]

His not being "sold on" carbon markets while supporting the implementation of REDD+ is reminiscent of concerns I addressed in the introduction and in chapter 1 of this book about the possibility of tying REDD+ explicitly to global carbon markets that never fully developed. In its iterations in Guyana and Suriname, REDD+ was supported through funds provided by foreign governments and interested parties. I assume that this should have assuaged the concerns of the UNDP-S representative to some extent because it decreased REDD+'s initial reliance on the market. Concerns about the means of funding REDD+ were also expressed by other stakeholders of its implementation. Some other consultants and policymakers tended to be publicly supportive of REDD+ while at the same time bemoaning the mechanism's reliance on bilateral and international funds, which they saw as able to provide only a limited level of financing.[13]

Li's fourth limit of (de)colonial mentality,[14] which recognizes the challenge of transforming political issues into technical areas of efficiency and sustainability, resonated in the critiques, expressed by some other stakeholders, of the idea that monetary values should be directly placed on nature. One representative of a civil society organization in Guyana explained that his concern about REDD+ was rooted in his view that when there is no longer a demand for something, it is usually gotten rid of. He wondered then what would be the effects of a decline in demand for carbon credits generated by forests.[15] This particular representative of civil society commented that whatever price is assigned to nature to facilitate its sale, "we can never buy nature back."[16] Further, he explained the fictitious nature of the valuation of the carbon sequestration work of the forests, which he recognized as not being real. He stated:

> Unless we recognize that it is not real, we are going to think that "oh yeah, we have got the best value for it," but we have completely misunderstood everything. I don't think market solutions are any solutions at all. We have really got to address the way we imagine our existence on Earth.[17]

In Suriname, representatives of nongovernmental organizations also voiced concerns about the inadequacy of monetary values for representing the worth of nature. A representative from Tropenbos International, an organization working towards increasing the scientific knowledge about the forests in support of conservation, explained that these numerical values are approximated and at times compared to the services nature provides in other countries. He explained that, as a consequence, the same ecosystem services in Suriname may be valued less than in other countries where there is a higher gross domestic product. Hence, he explained his view that while the "relative" nature of ascertaining the value of ecosystem functions may assist in planning efforts, the numerical values determined are not representative of a real value.[18] With this consideration in mind, he explained that Tropenbos is always careful to ensure that they do not use the word *payment* in their interactions with forested communities. They avoid this word to avoid giving the impression that communities will receive money because they reside in the forests. Instead, they use the term *benefit sharing* to suggest that the area could generate benefits based on the sharing of responsibilities.[19] These benefits could include development support, which would otherwise not be available. Interestingly, the representative of Tropenbos contextualized this argument by drawing attention to Suriname's history of receiving development aid, described in the previous chapter. He alluded to the corrupt behaviors of governing bodies that often result in the money not being used for intended purposes of development.[20]

Stakeholders involved in REDD+ implementation in Guyana were critical not just of the insufficiency of monetary values for representing some aspects of the environment, but of the inadequacy of the amounts received through the Guyana-Norway agreement. Another policymaker joined the chorus of stakeholders decrying inadequate valuation practices. He explained:

> What I see as the problem is lack of understanding of the value of nature, is lack of appreciation that there are other development paths to pursue, that removing forests for timber or conversion to agricultural lands or for mining is extremely short-sighted, because when you really look at the substratum, this is a very old geologic formation that supports extremely unique ecosystems found nowhere else on Earth. We have failed to appreciate the tremendous value of those ecosystems, the biodiversity and a number of goods and services that they produce. A valuation done in Guyana of US$580 million for standing forests, I think it is just a drop in the bucket.[21]

In these expressions of critique, policymakers and other stakeholders of REDD+ pointed to the insufficiency of monetary values for representing ecosystem services. The ascription of financial values to ecosystem services is effectively an attempt to make nature more visible to those governing it. In other words, the process of ascribing financial values to the work of the environment is generally pursued to make these services visible and legible to the market and those who make decisions according to market demands. Hence, critiques of this process in Guyana and Suriname highlight how this particular governing strategy, that of managing nature through financial abstractions, is challenged by some actors. The representation of ecosystem values through the market can be seen as an expression of neoliberal (de)colonial mentality, which seeks to inform behavior by incentivizing behavior change. Recalling Li's fourth limit of governing strategies, described above, the resistance expressed to this strategy through these critiques can be understood as the unintended consequences of governing efforts that seek to transform the politics of nature into areas of technical inquiry.

Challenging the efficacy and rationality of market approaches to managing nature even further, representatives of Amerindian organizations in Guyana pointed to the situation where the improper assignment of baseline deforestation levels led to controversy and the loss of revenue to the Guyanese state. In a report issued in collaboration with one such organization, Dooley and Griffiths described how the overestimation of the actual rates of deforestation in Guyana in the country's initial REDD+ agreement with Norway allowed for deforestation to continue its upward trajectory while allowing Guyana to continue receiving funds for avoiding deforestation.[22] Their work is worth quoting at length:

> Determining the rate of deforestation is a critical factor for determining performance based payments in reducing rates of forest loss. In the Guyana-Norway MoU (Memorandum of Understanding), the baseline rate of deforestation was set around twenty times higher than the actual historical rate of deforestation. During the first year of the agreement with Norway the actual rate of deforestation increased threefold (from 0.02% to 0.06% per year), yet Guyana received its first tranche of payments for reducing deforestation. This caused some controversy in international circles, and resulted in the MoU with Norway being modified to reduce payments if deforestation increases above 2010 levels, and halting payments if the deforestation rate exceeds 0.1%, which still allows for a considerable increase in deforestation. Deforestation has continued to rise since the Guyana-Norway MoU was put in place. The overall increase in deforestation compared to the last decade is due to the damage caused by gold miners, with mining remaining the main cause of deforestation in Guyana. The increased deforestation in 2012 could see Guyana lose as much as US$25 million, due to the modified MoU, which reduces payments if deforestation increases above 2010 levels.[23]

The process of implementing REDD+ in Guyana was fraught with methodological problems that did nothing to allay the concerns of some stakeholders. As described in the quote above, the Guyana-Norway REDD+ agreement was heavily criticized by bodies that observed that the agreement allowed for a tripling of deforestation levels, although this situation was subsequently rectified through the imposition of a more accurate deforestation baseline. However, the Amerindian People's Association (APA), who work towards representing the interests of Amerindians in Guyana, remained concerned about their perception that technocrats were benefiting the most from REDD+'s implementation. According to the APA, little benefit was accruing to indigenous communities while power was being consolidated in the hands of consultants and policymakers.[24]

Along with its insufficient or inaccurate ascription of financial values to nature, REDD+ was criticized by some stakeholders for overlooking the variety of unvalued ways (financially at least) in which environmental systems benefit human life and development. These tensions were perceived by some persons working in the area of environmental management in Guyana and Suriname. As one academic in Suriname saw it, the financial valuation of nature was necessary because "it is the only language that politics understands."[25] She argued that the use of biodiversity indexes in policymaking is futile because politicians are unable to connect to its values. She explained that "if you don't translate [your work] into economic values, your report, your PhD, or whatever will be something on a shelf. It is useless for the government to make the right decisions for natural resources."[26] Lamenting the perceived necessity of attributing financial values to environmental policy decision-making, she claimed that this was currently the only current way to make dialogue on conservation possible, as financial values dominate the decision-making language in policy circles.[27] She continued by referring to the gold mining activities taking place in Brownsweg, the location of the maroon captain referenced in the introduction. In reference to this, she recounted:

> The government allocated a small part of the park, and gold mining is heading towards the tourist area. There is a waterfall and above the waterfall, they are logging, and tourists are under the waterfall. Over the last few weeks, there were lots of discussion and meetings and they were trying to discuss where to put the miners, but the miners still got a small concession

> in the forests. They want to protect the park, but what do they get from the park besides the money from the gold miners? This is because they are not aware of the value of the park. The park has been assigned for ecotourism since the '80s and there are good management plans for ecotourism there, but when WWF started to pull back, the government no longer wanted to support ecotourism. So now the government says that tourism is down, so they allocate a piece of the mountain for gold mining. It's a perfect example of how they choose between conservation and that destructive development.[28]

While the allocation of a small gold mining concession in the park that had been designated for ecotourism does indeed demonstrate the tensions that exist in Suriname between those governing actors that prioritize conservation and those that prioritize economic growth, this particular occurrence demonstrates something more. It further demonstrates Li's pinpointing of the second limit to (de)colonial mentality. In this case, this expression of resistance and critique points to the inability of governing interventions to act as the sole determinant of outcomes. This is because the area referenced in the above quotation continues to be directly shaped by colonial histories traced in chapter 2. These histories resist governing interventions in support of REDD+.

In chapter 2, I highlighted that the Dutch colonial government, even prior to the area's designation as a site for ecotourism, had relocated Saramaka maroons to the Brownsweg area to facilitate the construction of a hydropower dam to power the then highly influential bauxite industry. Therefore, conservation priorities, in the form of ecotourism in this case, were unable to account for the colonial histories that had resulted in the maroons, who had been previously relocated to this area, subsequently turning to gold mining to sustain themselves. The eventual allocation of this area for ecotourism and the ensuing, separate inclusion of the work of all state forests of Suriname under the rubric of REDD+ conflicted, in these and other ways, with how people came to use the forests over time. These events are reminiscent of the observation of Simon Dalby,[29] who drew on the work of Christian Parenti to support his observation that "landscapes in those supposedly peripheral spaces have been remade by practices of colonization over the last few centuries."[30] Conservation and neoliberal conservation initiatives like REDD+ are not implemented on blank slates. Instead, these initiatives map onto land, forests, and use practices already shaped across centuries in particular ways. I will return to these tensions later in this chapter.

Attempts to fairly value nature are often complicated by the histories of the societies that have shaped that nature and that see themselves as intimately connected to it. While rooting the challenges to forest conservation in colonial histories, a representative of the Guiana Shield Facility of the UNDP explained that the cultures of Guyana and Suriname have not yet melded together. He stated that what their constituent groups have in common is a certain compatibility with nature. What has destroyed nature, he claims, is the tendency to respond to market demands in ways that allow for the destruction of the environment. He stated:

> In taking out and selling that gold, one, we destroy nature, which our cultures tell us is important to our livelihoods and important to who we are as a people. Two, the value we get for that gold is artificially decided. We don't decide that value. . . . It costs more than what the market says it costs because we are destroying other things to get it. We don't have the kind of mechanisms in place to ensure that receipts from that gold that we export are equitably distributed, that are leading to the development of our people, or that some of it has actually returned to reclaim the environment that has been disrupted. You know, we disrupt the environment. We disrupt the equilibrium.[31]

In this respect, it can also be questioned whether anticipated receipts from REDD+ would be distributed in a way that not only remedies the effects of these disruptions but that meets the needs of groups dependent on the environment for their sustenance and survival. Nonetheless, as this section demonstrates, the discursive transformation of the environment into natural capital furthered through REDD+ does not go unchallenged. People do in fact retain the possibility of acting differently. They often question the logic of REDD+ and the marketization of the environment it represents. However, the transformations intended by proponents of REDD+, combined with the continued, post-independence effort to govern Guyanese and Surinamese society in increasingly neoliberal ways, result in the fragmentation of the racialized subject introduced in chapter 3. I turn to the different levels at which this fragmentation is evident next.

Governmental Disparity

Policy incoherence, or as I call it, fragmentation, is evident at the state government level in policies to address gold mining. Fragmentation could be seen in Guyana when the Guyana Geology and Mines Commission (GGMC) was subsumed under the Ministry of Natural Resources and the Environment (MoNRE), itself established in 2011. This ministry was essentially given a mandate to simultaneously protect nature and exploit the resources it provides. In Suriname, this fragmentation was evident in the state government's engagement in REDD+ and its directive, according to a policymaker working in both countries, to exploit its natural resources to the maximum extent possible.[32] While some representatives of the state government of Suriname were harshly critical of the effects of gold mining and forestry on the interior environment of the country,[33] there were also reports of some government representatives engaging in mining by owning concessions.[34] However, the state government of Suriname appeared to recognize the damage to the environment caused by small-scale gold mining, which it sought to streamline within its borders. Since 2010, the Gold Sector Planning Commission (OGS) has attempted to create order in this area by working with some twenty to thirty thousand illegal gold miners to bring them into the country's formal economy. Previously, there were no legal mechanisms for facilitating gold mining in the country. However, due to the long period of uncontrolled gold mining, significant amounts of forests had been removed or degraded.

In Guyana, a similar duplicitous logic ensued where in the Low Carbon Development Strategy (LCDS), gold mining and its revenues and contribution to the local economy were often described by policymakers in glowing tones. These policymakers noted that mining and quarrying contributed to 10 percent of Guyana's GDP. In the space of one year (2011–12), the value of mineral production is said to have risen by 28.9 percent, to 175.8 billion Guyanese dollars. In terms of foreign exchange and exports, mining and quarrying rose from 58.7 percent to 61.6 percent of the total of US$1,291.1 million in 2012. This rise is due in part to the surge in the price of minerals, including bauxite and gold. Gold accounts for 78 percent of this output. Approximately twenty thousand persons were reported as being involved in mining and quarrying in 2012. The

LCDS described the likely increase of the output of gold as stemming largely from growing investments in large-scale gold mining.[35] Hence, while the state government of Guyana pursued REDD+, there was no associated effort to reduce mining. This will likely remain the case for as long as there is no change in the agreed, yet problematic, deforestation rate.

Fragmentation in how gold mining and REDD+ were viewed was also demonstrated in how miners were often vilified as sources of much environmental degradation and deforestation by government representatives and nongovernmental organizations but lauded for their contribution to the economy by other governmental officials and nongovernmental organizations. Yet, bodies that represent or regulate miners often portrayed them as innocent of claims that they were the most significant sources of deforestation. According to a leading figure in gold and diamond mining in Guyana, interviewed in 2014, there is no link between damage caused by the use of mercury and small-scale mining,[36] a connection known as irrefutable the world over. This leading figure also refuted the claim that gold mining is responsible for as much as 93 percent of deforestation in Guyana, referring to that as "nonsense."[37] According to my interview with the environmental department of the GGMC, which regulates mining in Guyana, this statistic is due to an error in calculation. The gold and diamond mining organization, on the other hand, attributed any other environmental damage to the inability of the GGMC to manage its miners and to enforce the mining laws.

However, according to media reports and an anonymous source, GGMC officers, while themselves forbidden from owning or participating in mining claims, do in fact own gold mines (even if unnamed on the mining claim establishing ownership). They are known to bend rules to suit their own operations and to accept large bribes, usually in the form of gold, from miners. Mining officers of the GGMC justified their lack of rigor by arguing that if they were to adhere to the formal rules of gold mining in Guyana, the industry would grind to a halt. The GGMC, however, agrees with the gold and diamond mining organization that attributing 93 percent of deforestation in Guyana to gold mining is an overstatement. According to their representatives, deforestation is partly the fault of the Amerindians who engage in slash-and-burn agriculture, which is incorrectly interpreted from the satellite data. GGMC officers also point to for-

Figure 10. Road to, and through, Mahdia, Guyana (Collins, 2014).

estry, and its clearing of forests to build roads to access timber resources, as a major source of deforestation and forest degradation.[38] This representation of mining as innocent of some of the charges against it is widely contested. Representatives of the World Wildlife Fund also attributed these burnt-out swathes of forests, visible from a flyover, to the activities of small-scale gold mining.

The GGMC also claimed that there is little opening of new mining areas since small- and medium-scale miners frequently rework old mining sites, especially when the viability of their enterprise strengthens due to increasing gold prices. When that occurs, finding a small amount of gold may bring in enough income to cover their costs and to provide them with a profit even in previously "worked" areas. This is because, as one miner in Mahdia, Guyana, explained, some gold always escapes the miner during the process. This particular miner claimed to have not removed any forest cover in his years working in an area, which, according to him, had been mined over five times.[39]

A miner in St. Luce, an area near Mahdia in Guyana, described:

> Well in some cases, people do rework the area. What happen, the kind of mining we do sometimes is throw away the gold right? So you would find

> some people working certain areas the old pork knockers were working back in '68 and time like that, because dredges and thing weren't so thing [popular] at the time. Is the Brazilians and so, I understand that brought in this engine kind of dredging thing, and the Guyanese run with it and, so you would find areas that have already been worked but because of machinery and engine and stuff, you won't find that the pork knockers, you know they late working [they would have missed it] would always find a piece of gold right there. You would always find an opportunity, so nuff [many] miners normally go to where them old pork knocker working, because they time was shovel and spade and drag-a-line.[40]

Clearly, at least to some degree, miners are responsive to market demands and external incentives as imagined through the logic of REDD+, especially in their self-sufficiency and demonstrations of self-interest.[41] The management of their operations is based on careful calculation of the economic benefits, and they are keenly aware of the input costs and the gold prices necessary to sustain their operations or make it profitable.[42] Their subjectivity is not straightforward though, since their allegiance to the mining enterprise is often unshakable. One miner replied, when asked about whether he will still be working in the Mahdia area in a few years, "I am not certain about around here, but what I am certain about is that I love this job and nothing is gonna stop me from doing it."[43]

The apparent inevitability of gold mining is also captured by the representative of Transparency International (Guyana) who works on issues of forest management. Referring to the then uproar over the fact that the Guyana government had lost US$20 million of the maximum US$250 million from its agreement with Norway due to increasing deforestation attributed mostly to gold mining, he described the government's response to the environmental threat:

> When the new minister of Natural Resources and the Environment took the bold and commendable step, . . . he said, "Oh we are going to put a hold on any new claims of river mining." Now river mining should be banned. Everybody knows that, even the people who are mining in the rivers know that it should be banned, but they don't want it to be banned because they're getting gold, they are getting money. But when he said he was going to put a hold on any new license, you're not saying that the ones that are there will have to stop, you're saying that you are not going to give them anything new.

Figure 11. Mined-out site in Mahdia, Guyana, showing the effects of gold mining on the forests (Collins, 2014).

> There was a huge uproar from within the mining sector, and within a very short period, he backed off. Why did he back off? Because it's gold! The miners will tell you that. The mining association will tell you that. It is gold. They have got the money.[44]

Largely, organizations and government officials have resigned themselves to the fact that mining will continue. Instead, they exercise their influence by working towards better management of the small-scale gold mining industry. While Guyana has already institutionalized gold mining and has instituted a formal system of managing it (challenges to its rigor notwithstanding), Suriname has relatively recently begun to institutionalize small-scale gold mining through the OGS. Little effort is made by policymakers to change gold mining trends, since this is unacceptable to key groups in society, some of whom argue that mining is central to continued national economic development, despite the arguments posited by other groups, particularly those who live in the forests, that mining is damaging to their prospects for health and well-being. Banning mining is officially a futile talking point, since this will never be accepted by key actors, especially when pushed by governments seen as unable to provide alternative income-earning options.

My interview with a representative of the OGS provided insights into how small-scale gold mining was viewed in Suriname, and into the state-led effort to manage miners and their forest use practices. He explained that in 2010, he was appointed to lead a special task force to bring law and order to the lawless gold mining sector, due in part to his law enforcement background. By the end of 2011, there were about 120 persons, mostly former military personnel, working to bring order to the interior. The aim of the force, he said, was to make miners become mainstream by "flipping" them, that is, making them get registered with the tax office and by helping them to become "bankable."[45] The miners were also being trained to use less mercury and to cause less pollution.

The OGS representative claimed to have since brought some order to the situation. The aim, according to him, is to treat miners like sick people who need treatment and to seek them out and make them orderly through medicine in the form of mining rights.[46] This action was spurred by the recognition of the government that gold mining is economically important for Suriname in the form of export earnings and jobs. This analogy of the miner as sick highlights their reputation as uncivilized threats to the well-being of nature and human life among some factions of the Surinamese society. Increased management through regulations and rules of disciplinary power are then needed to inculcate in these miners an ethic of environmental concern, in the furtherance of the aim of forest protection and REDD+.

The representative of the OGS compared this process of regulating small-scale gold miners to the setting up of a church in hell. He explained:

> Well, you are in hell, so everybody is bad, and you have to try to bring in these little souls and try to convert them. But still you are in hell so we will never forget that we are in hell because we are around these 4,000 gold mines, pits. People don't want to listen, they don't have time, they want to make fast money, they don't want to learn new skills, and they are surviving. That is the little church in hell, so the little church in hell tries to win souls and with the difference that my church has pastors with ATVs, and they go out.[47]

Through these and other efforts, miners themselves are depicted individually as a problem to be remedied through discipline. Yet, their collective identity oscillates between being represented as the troublesome drivers of deforestation of both countries and being considered the key drivers of economic growth. As a result, mining is treated in environmental policy circles in these two countries as somewhat inevitable, with governmental interventions being limited to reducing the harm of mining through management. The effectiveness of these harm reduction strategies remains an open question.

Policy efforts to have miners internalize the norms of better management of the forest resources complement the overall effort to govern the forests through incentives presented by REDD+. In response to efforts to discipline miners and their behavior in support of REDD+, gold miners make impassioned claims to having mined in areas for years and draw attention to their inability to earn a living through other sources, both during the colonial period and after independence. While their possibilities for earning might have moved from mining based on the foreclosure of other options to mining somewhat by choice, miners still claim to be marginalized by society and argue that they should thus be able to continue their small-scale gold mining, as it is essential to their survival and that of their families.

What I aim to have shown in recounting the incoherent, fragmented approaches to environmental policy landscapes in Guyana and Suriname in this section is how the difficulty of reducing the spread and intensity of gold mining resonates with Li's assertion, described earlier in this chapter, that governing is limited by the need to maintain some sort of societal equilibrium. As she explained, governing interventions often provoke and result in unpredictable outcomes and responses. Therefore, actors seeking to govern might be limited by their desire to maintain the delicate balance between social and economic processes.[48] As shown through the Brownsweg example and the strong allegiance of small-scale gold miners in Suriname to mining activity, gold mining presents a certain red line that cannot be crossed. Banning or reducing mining in support of forest conservation appears extremely unlikely or impossible, lest chaos be unleashed in society.

Group-Level Fragmentation

While they may have common origins, maroons in Suriname are by no means homogeneous in their views and political leanings. Even though several maroon communities are themselves engaged in gold mining, other maroon and indigenous communities are vehemently against mining in their areas. Opponents often point to the negative environmental and societal effects of the activity on communities where mining takes place and state their desire to keep it away from their areas.[49] These varying views of gold mining in maroon communities would be surprising only to those who maintain that forested communities are by nature better placed to protect the forests, as is common in discourses of the "ecologically noble savage."[50] Maroons' relationships with the forests, and their subjectivities in relation to them, are based on factors such as their historically racialized relations to the environment, experience with gold mining, proximity to the city, and whether they had been the victims of forced relocation to facilitate infrastructural development.

Miners themselves are conflicted in their views of mining. Some openly state that they destroy forests regularly as part of their work. They also accept that there are societal ills that come into communities through their work and behavior and that erode cultural norms and values. In the words of a maroon member of an organization representing the interests of miners in Suriname:

> Many times, you see the gold diggers looking for gold and destroying trees. Sometimes, you see the gold diggers, and by doing so, because they have their needs and they bring in prostitution coming into the community, something that destroys almost the system of how we know to live with each other.[51]

Conflicting approaches were also evident in how the miners viewed themselves in relation to the environment. Some small-scale miners believe that they can remedy their damage to the environment by being responsible, by filling the holes left in the earth, and by refusing to use poisonous chemicals.[52] These small-scale miners blame deforestation and environmental damage on large-scale mining companies.[53] This issue is debatable since the potential for remedial action after mining is dependent on the scale of the damage. In Suriname, researchers have found that small-scale gold mining and the repeated movement of the soil it entails slows regeneration and produces post-mining vegetation cover that is of poor quality relative to undisturbed forests.[54] These researchers estimate that mined-out sites will remain deforested for a minimum of ten years with perhaps centuries passing before mining pits bear any resemblance to old-growth forests.[55]

Another competing factor that impacts the behavior of gold miners is the gold price, which functions largely as the determining factor for whether a mining enterprise is viable. The gold price forms the measure against which input costs are calculated to estimate the risk of finding enough gold to cover these costs and make a profit. In 2014, when I gathered data for this project, the viable gold price was estimated at US$1,200 or US$1,300 per ounce of gold.[56] As the former leader of an association of gold miners in Guyana explained, regarding the rise of gold to economic centrality in Guyana:

> Up until 1972, 1973, miners were mining diamonds, diamonds was what people mined. Gold was by the wayside. If you find some gold, you make Jewelry; you give it to your girlfriend or somebody, because of the price of gold. The price of gold those days were in the London market. The price of gold was fixed. It was 33 dollars an ounce, period. It was fixed by the US government. An incident came up, I think, between the US government and the French government, where the French government demanded payment in gold rather than the francs that the US had holding. So, the Nixon government, it came off of that, the 33 dollars an ounce. So, the price of gold moved dramatically. As a matter of fact, by 1978, it was over 800 USD per ounce. In those days, Guyana money was pretty decent. At 800 dollars per ounce, we were getting 1,200 Guyana dollars, so what happened, a lot of people moved from diamonds right away to gold. That was when the gold thing started.[57]

Diamonds are not a main income source in the Guyanese economy. Gold shot up in importance and earning potential and has since remained central.[58] For the gold miners, despite their awareness and behavioral responsiveness to the gold price, small-scale gold miners see mining and their residence in forested areas as a fixed, central part of their identity, as exemplified by the maroon captain with whom I opened this book.

Other actors related to REDD+ in Guyana and Suriname are also highly conflicted in the way they describe and relate to the forests, leading to political tensions. Many forest-dependent communities challenge the expansion of capitalist activity to their communities, as well as their perceived inability to share in the benefits. Some community members bemoaned their limited access to employment opportunities in extractive endeavors. Specifically, they alleged that companies come to their communities, extract the resources on which they depend, and bring workers from their countries, largely Brazil and China, limiting local community access to employment. Some community members further alleged that these infractions on lands they consider their own are made possible through covert deals between the state government and the foreign companies. Amerindian community members with whom I spoke expressed their concern for the forests within which they live, both because of the damaging effects of extractive activity on the environments on which they depend, and also due to their perception that they were not being given their fair share of the proceeds.

Even within the collective effort of indigenous groups and maroons in Suriname to work together on having their land claims recognized, there are differences between these two groups in how they approach the issue. The Saramaka maroons, for example, took their case for land rights to the Inter-American Commission on Human Rights (IACHR) of the Organization of American States (OAS), while indigenous groups worked collaboratively through a civil organization that represents all but one indigenous community in Suriname. Forest communities are by no means a cohesive entity. Within the policy data, indigenous groups are overwhelmingly portrayed as being more concerned for the well-being of the environment, considering especially that they have lived within these forests for generations, even before the arrival of the colonizers and other ethnic groups. Maroons, however, are depicted in policy documents as people forced into the forests by extenuating circumstances. Although they are recognized as having formed bonds with nature and the forests since their relocation, their relationship with nature is seen to be a few shades more exploitative than that of the indigenous groups. While the damaging forest use practices of some indigenous community members were acknowledged by some of the people I interviewed,[59] indigenous groups continue to be broadly depicted in policy documents as harmless forest dwellers who have been able to manage their natural resources without depleting or worsening them.[60] In fact, it is often suggested that their knowledge of the environment should be drawn on for solutions to environmental challenges.[61]

Internal Fragmentation

Fragmented views of REDD+ and forest conservation do not stop at the collectively identified subjects described above. The maroon captain, with whose interview I opened this book, further exemplifies the fragmentation of individual subjects. I provide more detail from his interview here to further highlight the internal conflicts affecting behavior patterns on the ground:

ME: Can you tell me a little about yourself and what you do?
MAROON CAPTAIN: I am the captain of this village. I am a gold miner.
(...)
ME: Do you think the forests should be protected?
MAROON CAPTAIN: Yes.
ME: Why should the forests be protected?
MAROON CAPTAIN: For the future.
(...)
ME: Who do you think should protect the forests?
MAROON CAPTAIN: The government.
ME: Why is it important that they should protect the forests?
MAROON CAPTAIN: The government has the power in Suriname, that's why.
ME: What activities cause trees to be cut down?
MAROON CAPTAIN: Gold mining, quarrying, logging and making products out of the wood.
ME: Do you think you damage the forests?
MAROON CAPTAIN: Yes, I think so.[62]

As previously described, the satellite communities of Brownsweg, where this interview took place, were designated in Surname as transmigratory because of the colonial government's stated intent, at the time of internal relocation, to move the communities to more permanent areas of residence at a later date. Trapped in a state of limbo since the 1960s, members of this community turned to small-scale gold mining to sustain themselves. By no coincidence, it is this group of miners who was granted permission to mine for gold in a neighboring park. As previously mentioned, this park had been designated for ecotourism, but after consulting

Figure 12. Effects of gold mining in Brownsberg nature park, Suriname (Collins, 2014).

with the miners, the head of the OGS decided that the miners would be allowed to continue mining there. The head of the OGS stated that given the injustices meted out to these communities over centuries both before and after independence, the miners would not be denied the opportunity to continue the mining they had already started in the park.

The difficulty of conserving the Brownsberg nature park in Suriname illustrates not only the internal conflict of the maroon captain, but also the difficulty of reconciling the need for economic growth on the part of the Surinamese government, and the colonial history and reshaping of the area. The authorities of the nature park wanted the miners out of the park, but the representative of the OGS stated that "this, of course, was not going to happen."[63] He explained that after extensive consultation with the three communities that mine near the park, the OGS decided that since these people had been mining there for twelve years already and had caused significant destruction of the environment in that location, they would be allowed to continue mining there, considering that the location had enough gold reserves to keep the miners engaged for an additional ten years. In order to ensure that the mining does not spread to the remaining areas of the park, an outpost was created in the buffer zone to ensure that the miners stayed in their designated area.[64] These communities also have difficulty finding land for mining and are, as previously mentioned, often expelled from their lands when large-scale mining concessions are granted.[65] Unaddressed colonial histories are palpable in this case, directly contributing to the challenge of protecting forests in Suriname, since these groups cannot in good conscience be prevented from engaging in this means of earning income given that their precarious situation is the result of unremedied historical injustices, a connection they themselves express.

CONCLUSION

As shown in this chapter, the actors that benefited and continue to benefit from deforesting and environmentally degrading activity in Guyana and Suriname are not limited to colonial centers in Europe. In both countries, the need for decolonizing environmental governance is evident in how the very market-oriented, extractive logic and system implemented by the colonizers has taken root and continues to strongly influence the response of gold miners to market signals.

REDD+ governs in an imagined view of the rational subject of *homo economicus,* which seems almost entirely absent among REDD+ stakeholders in Guyana and Suriname except in the case of small-scale gold miners. While all actors related to REDD+ responded to financial incentives to varying degrees, only gold miners did so directly and primarily. Nevertheless, gold miners see small-scale mining as a way of life. Hence, the general absence of amenable incentive-oriented, self-governing subjects in Guyana and Suriname and the strong allegiance of these subjects to gold mining where they do exist limits the ability of REDD+ to incentivize behavior away from gold mining. However, this chapter demonstrates that, especially in the case of gold miners, undisciplining the disciplined subject comes with inherent risks. Through disciplinary measures gold miners, who are known to be the largest threats to the forests of Guyana and Suriname, were brought under varying degrees of state control and management. This discipline has therefore lessened the likelihood of their wantonly felling trees, polluting waterways, and degrading the environment at will. Put simply, discipline brought some level of control to the problem of deforestation on the ground. In the case of gold miners, efforts to undiscipline their behavior might indeed walk back progress that has been made on limiting harmful environmental impacts.

In relation to some other actors identified as significant for REDD+ in these two countries, undiscipline in the interest of the environment would be desirable. The benefits of undiscipline are evident in how some actors question and challenge state-sanctioned marketization and the expanding use of markets for determining environmental affairs. Undisciplining those subjects of colonial governance who do not question the tendency to center power in the state, the frequent granting of concessions in lands claimed and used by forested communities, or the expansion of markets to ordering all other social concerns might have beneficial impacts for nature, conservation, and climate change.

The undisciplined subject does not automatically defer to markets. If ways of being other than those that operated in support of the colonial enterprise are prioritized, avenues may open up even for small-scale gold miners who have turned to gold mining, to some extent, through a lack of other means of sustaining themselves. Focusing on gold mining itself may encourage a reliance on instilling further discipline in the interest of conserving the environment. However, broadening the temporality of the decolonial horizon to consider the historical injustices that brought some miners to rely on mining for sustenance in the first place may open opportunities for the undisciplined miner himself to envision other ways of being. This is especially likely if this is done while building on the need to decenter the market, as detailed in chapter 3.

In any case, the circumstances of the deforesting gold miner and the need to discipline or undiscipline him or her demonstrate that decolonization comes with immediate advantages and/or distant trade-offs because modernity-defining histories cannot simply be walked back to a time of unbridled opportunity. These trade-offs, once acknowledged, can be grappled with in ways that do not default to the unquestioned pursuit of development and modernity, but that open spaces for other ways of being to shine through. Hence, taking these openings of undiscipline seriously forms the third step towards decolonizing environmental governance I offered in this chapter. Accordingly, the fourth step, detailed next, would see discipline countered with truths and their ideas operationalized.

5 Counter Discipline with Truths

In May 2014, the state government of Suriname hosted the Highly Forested Low Deforestation (HFLD) Conference. The aim of the conference was to make REDD+ clearer to local stakeholders while sharing country-level experiences of preparing for REDD+ at the intergovernmental level. The conference brought together national representatives of Guyana, Suriname, Nepal, Bhutan, and Belize; representatives of international conservation organizations such as Conservation International (CI) and the World Wildlife Fund (WWF); and officials from REDD+ funding organizations such as the World Bank and the United Nations REDD+ Programme (UN-REDD+). Representatives of civil society and forested communities also took part.

During the proceedings, a representative of a maroon community stood up and argued vehemently that the root of the problem of deforestation and forest degradation in Suriname was that the state government of Suriname grants concessions to companies while not allowing communities to use the forest. He asked:

> What is the world doing? REDD+ is about protections in Europe. Villages and houses are given to multinationals for mining. How does the world fix

> this? How can traditional practices continue in the face of these challenges? They cannot fish, hunt, or plant without permission from [the government]. People are forcibly removed from concessions. People were moved to new spaces and they are being moved even from that. Maroons moved to the cities and will engage too in environmentally degrading activities. They don't care that our ancestors are buried there. Then when we adopt their way, we are stopped from that too.[1]

The maroon representative cast REDD+ as simply the latest installment in a tradition of government policies that burden forest communities. He drew on his childhood and the challenges his ancestors historically faced in order to position maroons' communal attachment to the land in opposition to demands emanating from the head table that the communities also work towards avoiding deforestation. The head table, composed of representatives from international organizations and other state governments, responded to the maroons' challenges by relegating the maroons' concerns to the internal affairs of each country, and hence as outside their immediate purview.

The representatives of forested communities present at the conference remained unswayed. One retorted, "There are continued claims for indigenous rights and the needs of forests seem to be taking priority over the rights of forest peoples. Maybe we should become animals and you will protect me?"[2] During the conference, the forest community representatives frequently vented their frustrations in this manner. In so doing, they clearly demonstrated how historical injustices and racialized relations to the forests featured in national discussions on the implementation of REDD+ in Guyana and Suriname. The frustrations expressed by community representatives showed how these histories and relations continue to permeate and form sticking points in debates on REDD+ and forest governance. To a large extent, these expressions remained rooted in the relative, historical autonomy that forested communities once had over the forests; as community representatives, they challenged the state governments' sovereign approach to managing land they thought to be their own.

• • • • •

The chapters of this book have explored, thus far, some of the different threads through which colonial histories continue to bind forest governance in the Amazonian Guiana Shield to particular pathways. The power imbued in the state government to legally manage the forests, referred to as sovereign (de)colonial mentality, forms the first such thread. The increasing deferral to markets as primary adjudicator of environmental management decisions, referred to as neoliberal (de)colonial mentality and evident to a greater degree in Guyana, represents the second. As explained in the introduction of this book, these expressions of coloniality are also strategies for decolonization, hence the use of the term *(de)colonial.* However, these colonially rooted strategies are resisted, in various ways, by groups identified as stakeholders of REDD+ in both countries. Meanwhile, REDD+ proponents imagine these stakeholders as subjects to be governed who are responsive to market concerns. Where stakeholders fail to demonstrate the characteristics of rational and responsive neoliberal subjects, REDD+ proponents, particularly state governments, work to discipline these subjects into being. Imagined spatially, the threads that bind forest governance to particular pathways spin from the historic and embryonic colonial formation of the Guyanese and Surinamese states inwards to different groups at the subnational levels and come to rest in the forests with the communities still residing there.

Decolonizing forest governance depends on making these truths visible. I understand *truths,* drawing on Foucault's governmentality framework, as referring to culturally informed understandings of the world that were marginalized or outright negated during colonialism. Truths, in this sense, point to those strategies or mentalities of government that function according to religious texts, revelation, and interpretations of the natural order of the world.[3] Contrary to neoliberal, sovereign, and even disciplinary (de)colonial mentalities that depend on coherent governance from above, truth (de)colonial mentality retains the ability to capture more grounded and decentralized ways of relating to the environment, such as those employed by the people confronted by Europeans and by those brought to the Guiana Shield during colonization.

The events that shaped the forests of Guyana and Suriname following the historical confrontation between European explorers and people indigenous to the Guiana Shield represented an initial clash of truths. These truths were informed by different approaches to ordering the world, of valuing human and nonhuman life, and of being in the natural environment. Those modes of relating to the environment that were overtaken in this colonial encounter took place in accordance with certain preordained precepts as set out, for example, in traditional African spiritual practices, indigenous epistemologies, Hinduism, and other non-Eurocentric ways of interpreting the role and significance of humans in their environments and the world. Despite their marginalization during colonialism, these truths continue to challenge subsequent sovereign, disciplinary, and even neoliberal attempts to govern the forests of the Guiana Shield after independence. These truths naturally also challenge the pursuit of forest governance through REDD+.

Indigenous, maroon, and other populations forcibly relocated to the Guiana Shield brought with them diverse ontologies and truths that were underpinned by an acute awareness of the dependent and interconnected nature of the human and the environment.[4] Throughout slavery, indentureship, independence, and subsequent post-colonial efforts to meld multiethnic societies in Guyana and Suriname, these varied ways of relating to the environment were sidelined by Europeans, who had instituted their own norms of viewing the environment. Those norms tended to mediate the relationship between humans and the environment through the lens of the market and its principles in the aim of accumulating capital. Hence, while capital was becoming central to the organization of social relations in Europe, in part by constructing a new patriarchal order that disenfranchised women,[5] practices of controlling and disciplining populations in the interest of capital accumulation were being perfected in the colonies. Those practices recast the environment as useful and valuable only insofar as it directly supported human endeavor through the market. Despite their marginalization and suppression, however, alternative ways of being in nature continue to exist. These varied expressions of these truth (de)colonial mentalities can be seen, for example, in non-Eurocentric religions, indigenous cosmologies, and culturally informed traditions and myths that still shape the social life of both Guyana and Suriname.

State governments of both Guyana and Suriname recognize the fact that forested communities often see themselves and their well-being as interconnected with that of the forests. State governments then use this recognition as evidence that these communities should be categorized and treated as special in the norms, laws, and societies of Guyana and

Suriname. During colonialism and even after independence, state governments of both countries worked towards integrating and disciplining the forested communities into active participation in the national economies and development trajectories centered on the coasts. Through REDD+, however, forest communities that had previously been left out of the national thrust towards market centeredness are now being further integrated into the market and the social expectations this integration necessitates. This takes place, in part, through the neoliberalization of forest conservation.[6]

This chapter shows how the resistance of forested communities to REDD+ succeeded in impeding the initiative's implementation. It also shows how REDD+ proponents respond by targeting these communities and seeking to integrate them into forest governance practices as eyes on the ground in support of sovereign forest management. Through these and other state-led efforts, communities are being integrated into conservation initiatives aimed at better managing the forests they have lived in for generations. This forms an interesting proposition, since the timber and gold mining concessions that historically contributed to deforestation in these places are often undertaken with the explicit approval of the state governments, often without regard to community interests. In these ways, international demands in the post-independence period for the conservation of the forests, which stood witness throughout the colonization and eventual creation of Guyana and Suriname, interact with the traditions of indigenous and maroon communities who live in and with the natural environment in not quite capitalist ways.

In this chapter, I trace some of the state-led governing efforts undertaken to discipline forest communities into REDD+'s overarching aim of keeping the forests standing while making them visible and financially valuable. Discipline becomes a necessary tool for REDD+ proponents here because the truths held by these communities have marked them as different and outside of REDD+'s market logic. These truths are often expressed in communities' rhetoric and lived experience as relics of a somewhat incomplete process of colonial subject formation. Efforts to integrate these communities into the REDD+ effort proceed accordingly. These efforts are based on latent assumptions that the development and market-based activities centered on the coasts exemplify the standard, taken-for-granted model of progress, and that forest-dependent communities need to be brought up to speed with their coastal counterparts after having been left behind.

As explained in chapter 2, comparisons of Guyana and Suriname with European former colonial centers support arguments for the industrial development of the former. Similarly, comparisons of interior forest locations with the more developed coastal areas justify the need for the assimilation of the forest communities and their resources into national development efforts. In view of the linearity of the national drive towards modernity, this chapter explores countering discipline with truths as a strategy in the quest to decolonize forest governance through a focus on the long-standing issue of land rights. In essence, therefore, this chapter shows how forest-dependent people are integrated into REDD+ forest governance through disciplinary methods that further fragment their racialized subjectivities.

TRUTH AND DISCIPLINE IN GUYANA

The truth that indigenous communities have special, spiritual relations with forests is evident in national policy documents in Guyana that relate to REDD+.[7] Amerindians' significant claim on the territory of Guyana is rooted in these collectively held truths. The state government, in recognition of indigenous demands for land rights, agreed to have these issues addressed through the Amerindian Development Fund (ADF), established in 2015. This project was funded by the Guyana REDD+ Investment Fund (GRIF), which received proceeds from Guyana's REDD+ agreement with Norway. The project document of the ADF represents an archetypal example of Guyana's state government's attempts to govern indigenous communities in accordance with the logic of the market. Hosted by the United Nations Development Programme in Guyana (UNDP-G), the ADF's project document described Amerindian people as "at varying stages of integration with the national economy,"[8] defining them as cultural groups that generally feature well-preserved traditional lifestyles. The aim of the project was to gradually integrate these communities into the "production and consumption structures of the national economy."[9]

The ADF holds as its primary focus a problematization of poverty in rural communities in Guyana. While it is by no means the first policy intervention to do so, it does so specifically in the confines of REDD+.

However, poverty in the context of forested communities is not simply a lack of cash, but the absence of a means of survival.[10] Amerindian communities have been living in the forests for at least centuries, well before the colonial enterprise and the subsequent national move towards increasing reliance on the market. In modern-day Guyana, some communities have titles to their land and the aboveground resources on that land. They live off the land and depend, in some cases, on solar panels for electricity and on lakes for water while growing their own food. While indigenous communities do still commonly need financial resources, especially since some of their members frequently travel to the coastal cities, the communities are often less dependent on cash for their day-to-day survival than their coastal counterparts are. Poverty in this context is an outcome of engagement with the increasing turn towards capitalist means of production.[11] Understood in this way, poverty in indigenous communities can be recognized as being worsened by activities like gold mining and logging that pollute the rivers and chase away animals that community members hunt for food. In that sense, these market-oriented activities undermine the capacity of the communities to sustain themselves outside the market. This interpretation of poverty, however, runs counter to the interests of state governments in continuing to access natural resources on lands claimed by forest communities. As a result, the histories of state formation that saw indigenous communities impoverished go unaddressed in the plans and interventions of the ADF.

The project document establishing the ADF commences with a problematization and role definition that sidelines these subsistence considerations. In its situational analysis, the project document marks out the problem it intends to address, namely "the situation of poverty in the rural interior—where most Amerindian Communities are concentrated,"[12] and the "specific limitations associated with developing rural economies in the context of the remoteness of Amerindian communities and the nuances of doing so in Guyana."[13] The document attributes this poverty to the physical conditions of the environments in which communities live and to the absence of infrastructure linking them to capital that could enhance their business opportunities. In hopes of remedying these challenges, the ADF aimed to support Amerindian communities in the following sectors: processing, village infrastructure, tourism, manufacturing, village business enterprise, and transportation. Support was to come through the provision of financial resources and the strengthening of implementation capacities for sustainable livelihood activities.[14]

These sectors were identified as beneficial to the socioeconomic development of Amerindian communities in Guyana in each community's Community Development Plan (CDP). The CDPs were said on the website to have been developed democratically by the villages, even though this has been reported to be false by one well-positioned interviewee. Instead, he claimed that the twenty CDPs made available to me were in fact written by the staff of the Amerindian Affairs Ministry in Guyana.[15] Some communities were represented as requesting improved machinery, such as the provision of an all-terrain vehicle to improve employment opportunities by offering rides for tourists. Other communities were represented as desirous of increased income through activities like honey production, which was facilitated through training in financial and business management. The underlying logic of the CDPs is illustrated through the proliferation of quotations, such as the following, that were explicitly growth oriented:

> Economic health refers to the need to strike the balance between the costs and benefits of economic activity, within the confines of the carrying capacity of the environment. Based on the financial calculations and with an expected Financial Internal Rate of Return (FIRR) of 7.25%, giving the risk, and in spite of the 3 years duration to cover the cost of capital, the proposed project will still yield positive economic benefits for the Village with spin off effects for community economic development, labor market development, infrastructure and agriculture with minimal impact on the environment.[16]

There were frequent references in the CDPs to the need for technological innovation but never a reference to the need for the clarification of land rights or to the disadvantaged position of the communities in relation to gold miners operating on their land. In an argument reminiscent of the claim that REDD+ is performing communities,[17] it was clear that the

CDPs were, at least, guided by government authorities. They were presented in the same template with the same depiction of technology and the market as the saviors of indigenous communities through access and capacity building. The CDPs reflected the ways in which the state government of Guyana wanted indigenous subjects to be depicted and subsequently disciplined.

For example, an ADF project document states:

> Due to terrain and other natural conditions, critical infrastructural challenges confront the feasibility and sustainability of many business ventures pursued by Amerindian Villages and Communities. Accessibility and logistics: remoteness and costs associated with transportation, communication and other logistical needs contribute significantly to the loss of competitiveness and necessitate the procurement of technical and skilled services, which are vital in ensuring business viability.[18]

In its focus on access and competitiveness as contributors to Amerindian poverty, the project document removes from view the partially unsatisfied claims for land rights these groups have asserted over centuries. It does so while justifying infrastructural development as a means of facilitating economic growth to lift the groups out of income poverty. A tangential consideration is that these documents do not mention that increased access to the markets in the cities would also result in increased reciprocal access to the communities by city dwellers, a possibility that comes, for example, with the risk of reducing the presence of animals the communities hunt for subsistence and sale. In its vision of Amerindians and their living conditions, the ADF identified as necessary remedies for the situation of poverty in Amerindian communities the following: the provision of roads and cheaper transportation, improved communication, and the education of community members on the means of competitiveness and competition. These remedies facilitate an overt policy effort that attempts to transform Amerindians into self-governing subjects that are then more amenable to being governed according to market rationalities, and hence to REDD+.

The very nature of Amerindian communities themselves is problematized in the project document, which states:

> By nature, Amerindian communities evolve and reform through community ownership, responsibility, volunteerism and communal labour. As such,

> there is a significant gap between livelihood needs and economic enterprises that requires careful and attentive support including nurturing, monitoring and rapid response troubleshooting.[19]

Here, community members are naturalized as communal and painted as in need of reform if they are to become amenable to market governance. It remains an open question whether the effort to make these communities more amenable to business ventures and self-management requires a change in the very nature of the communities that the document identifies as the root of the problem.

The project document is underpinned by the view that access to markets and behavior suitable for the management of economic enterprises can ameliorate poverty in Amerindian communities. While the ADF claims to have an interest in strengthening the management of internal village structures, it does so because its architects see this factor as critical to the implementation of the CDPs. The ADF further posits that increased access to market information and communication technologies are the answer to the challenges of indigenous communities. Access to financing and investment options and an adequate supply of green energy are also commonly framed as potential solutions.

The market orientation of the project document is made explicit in quotations such as this:

> The distribution of a micro capital grant for business development to Amerindian communities cannot by itself be the panacea for socio-economic development, though it does have the potential to stimulate economic activities further. The project aims to establish market access and improve linkages with players in the private sector. The task at hand therefore is to address specific capacities and capabilities related to communal business development and management for implementation; negotiations and bargaining with private enterprise; information asymmetries; and market integration.[20]

The recognition that capital grants are unable to function as a magic bullet to solve poverty in the communities is the only reference the project document makes to circumstances beyond its immediate control, an issue on which it does not elaborate. It does continue, however, to be apolitical in its stance and in its presentation of technocratic solutions for the deeply entrenched problem of viewing indigenous communities as relevant just when their lands are needed to fuel industry and economic growth. The ADF project document blurs out the historical events and power relations that have moved indigenous communities deeper into the forests and limited their livelihoods, as outlined in chapter 2.

These solutions to Amerindian poverty conveniently ignore the root causes of the poverty it identifies. As I see it, communities became discursively poor when measured by increasingly dominant market-based approaches to valuing human-nature interactions.[21] Considering that these communities have lived for centuries in the forests of the Guiana Shield, modern-day poverty is a comparatively recent categorization. If one looks far back enough, it becomes clear that poverty understood in this way emerged in these communities through repeated attempts to integrate them into market activities centered in the coastal areas. Communities are then kept poor through environmentally exploitative activities that undermine their ability to sustain themselves, activities that are sanctioned by the subsequent governments of the modern states of Guyana and Suriname. Hence, income poverty is constructed as the communities increasingly become subsumed into the capitalist paradigm, as their environments are valued only when exploited and their ways of life viewed as impediments to that exploitation. This situation is made possible through their continued interactions with the capital centers of both countries, whose leaders imagine forest dwellers to be backward and different and regard their environment as resources to be drawn upon to fuel national and foreign exploitation and economic growth.

Further, this situation is made possible through the marginalization of epistemological indigenous truths and the establishment in their place of capitalist precepts. In the post-colonial period this exploitation continues to be supported by an absence of adequate rights to the land, to varying degrees in each country, and increasingly, through the communities' own exploitative pursuits in the aim of attaining development and wealth. The UNDP-G, through the ADF, supports this latest endeavor, utilizing the solutions readily available in its arsenal to suggest improvement for the communities while ignoring the more entrenched and fundamental histories of disempowerment that left communities in disadvantaged positions, echoing concerns expressed by anthropologist James Ferguson on anti-politics.[22]

Some indigenous community representatives do vocally protest these events. In 2014, a special report on REDD+ produced by the Amerindian People's Association (APA) and the Forest People's Program detailed how the Guyana Forestry Commission routinely grants concessions to logging companies in lands that indigenous communities consider their own without community consent. The special report recommended extensive structural changes to Guyana's governing architecture. These included legal and regulatory reforms to the sovereign expressions of (de)colonial mentality in accordance with international obligations that require them to consult with indigenous communities before granting concessions, the adoption of a fair and transparent process for resolving land conflicts, and the annulment of mineral and lumber rights issued to third parties that are currently in effect on customary land without community consent.[23] The APA representatives explained their view that while REDD+ may indeed be useful for protecting the forests and might indeed have beneficial outcomes for Amerindian people, they remain concerned about its execution, particularly in relation to land rights that they feel must be strengthened before they lend their support. They further pointed to the "inadequacy of the Amerindian Act to fully protect indigenous land rights."[24]

The Amerindian Act of 2006 to which the APA refers serves as the legal basis on which the right to the lands claimed by Amerindian communities was established. It was presented by the state government of Guyana as a sign of their willingness to engage with indigenous concerns and to recognize the Amerindians' need to have some degree of autonomy. It is, however, often challenged by some representatives of indigenous groups in Guyana as insufficient. Behind these expressions of frustration lies an awareness of historical indigenous use of the land prior to the European contact on which the state is built. For example, lamenting the constitution's retention of rights to mineral resources, APA representatives argued enthusiastically against the granting of licenses on traditional lands, explaining:

> You have areas that fall within traditional lands that are not titled Amerindian communities that will be considered under the REDD+ scheme. At the same time, you have, and that's competing within the indigenous traditional interests, you have indigenous communities also competing and conflicting with the mining and the forest sectors, and protected areas, and

> who else knows, or what other sector as well. That's what the Amerindian Act says, that villages have to give permission. But the constitution retains the rights to mineral resources, and the mining act takes precedence and all resources belong, and that is what the agency has been promoting, that is what the courts have upheld. Because the Mines Commission has the right to grant permits prior to a community giving permission. So, the miner can easily have a piece of paper which says, "I have permission to do this mining here," then move to the community and say, "Can we have a mining permit?" In the case of traditional lands, that doesn't apply. The Mines Commission will give a mining permit, and that's why you have one community being taken to court over and over and over again because they are seeking to protect their waterways, they are seeking to protect their lands. One of the only untouched rivers in that area and they [communities] are seeking to protect it, well the Mines Commission is saying the miners have the right [to mine there]. The court is saying the miners have the right. . . . This is where the clash between protection of the environment and extractive activity and protection of a people comes into play.[25]

In the language I draw on in this book, I would say instead that this is where the clash between sovereign and truth (de)colonial mentality "comes into play." In further challenging sovereign expressions of (de)colonial mentality around land rights, the APA explained that communities often use borrowed language from the state government that leads them to refer to their rights through "a piece of paper that the government gave to them,"[26] but if communities are asked more simply about where they hunt, fish, and reside, a more comprehensive area of land would be identified as theirs.[27] As one indigenous community member explained at a REDD+ consultation meeting, "Amerindians have lived in the forest for years and have always protected it. People can't come to tell me that I should stop my way of life because it wouldn't be fair."[28]

These disputes are based on what the APA sees as discriminatory norms with roots in the colonial period that continue to be manifested in Guyana's national legal framework. These norms support the idea that all of the untitled land in the country is solely the domain of the state. This position does not consider indigenous peoples' customary land ownership rights as automatically valid and is based on the understanding that the rights to the lands of indigenous people were voided by the colonizers from which the state emerged.[29] Hence, property rights are the sole domain of the state and must be defined by their acquiescence or granting of these rights.

Thus, the traditional use of Amerindian land must be granted by the state in order for the community's use of that land to gain official recognition. As a result, many settlements across the country are entirely without title. The ability to log and mine within the communities is at times also excluded from the community's land title, and local court rulings tend to support these extractive activities carried out by noncommunity members when they are challenged by the communities.[30] The APA sees this as being in contravention to international human rights proclamations, especially when one considers that forestry and mining concessions are often issued by Guyana state government bodies within the areas considered to be customary lands by the Amerindians. Consequently, conflicts between the communities, loggers, and miners frequently occur.

According to the Amerindian Act, titled lands are owned and managed by the communities, who have complete rights to the land except for mineral rights. By law, they can veto attempts to have medium- and small-scale mining take place on their land, but not large-scale mining.[31] According to the Guyana National Land Use Plan, land lease types can be small-scale (27.58 acres) permits, medium-scale (150–1,200 acres), or large-scale prospecting licenses (500–12,800 acres) that are renewable yearly and can be converted into mining licenses (up to 20 years or life of deposit, renewable).[32] Public land is, in effect, all land that is not owned privately or by Amerindian communities. The Guyana Geology and Mines Commission (GGMC) and the Guyana Forestry Commission (GFC) administer leases for resources on and under the land, for mining and forestry respectively, with each agency imbued with the capacity to issue titles for different purposes over the same land space as discussed above. Clear land tenure regimes are central to the REDD+ mechanism and to the process of demarcating forests that allow for carbon to be created as a commodity that can be sold and traded on the market.

Recounting the beginning phases of REDD+ readiness preparations in Guyana, the APA described how an early draft of the readiness preparation proposal (R-PP) was approved by the Forest Carbon Partnership Facility of the World Bank in June 2009 despite what they describe as poor participation of the indigenous peoples. Although some corrections to the R-PP were requested and carried out subsequently, some issues remained unresolved while preparation for REDD+ forged ahead.[33] Meanwhile, Amerindian communities continue to face the unwanted effects of concessions, often granted to foreign companies, by the state government in lands they consider their own. On the other hand, the representative body of gold miners in Guyana criticized the Amerindian effort to gain recognition of their land use by stating that the Amerindians in Guyana have interests solely in lands known to have gold. These conflicts are actively playing out in the law courts of Guyana, with Amerindian communities usually suffering grave losses based on the fact that some mining licenses were given out prior to the signing into law of the Amerindian Act that gave them some rights over their resources.[34]

Detailing the fight over land rights in Guyana, the APA explained that the lands communities expected to be granted to them are at the same time being granted to others as mining and logging concessions. Amerindians have sought to have these perceived injustices remedied by the courts but have lost their cases and subsequent appeals. This was illustrated by a court case in 2013 where the GGMC asserted that it would not be accepting fees for mining concessions in proposed Amerindian lands but was forced to continue doing so because the miners filed an injunction and won. The miners' victory was based on the fact that the mining claim was granted before the award of the Amerindian land title. Nevertheless, according to the National Land Use Plan, cost is the major impediment to the country's progress in demarcating land. The plan also reflects the government's commitment to clearing all outstanding requests by 2015, through its partnership with the UNDP, a target that was not met.[35] Some of the funding the country should gain from the REDD+ process is allocated to continuing this demarcation process.[36]

Forested communities and their representatives are essential actors in REDD+ preparation and implementation because of requirements of the World Bank and United Nations that these communities be consulted before their forests are included in the mechanism and because some communities hold legal titles to land. That, along with the physical proximity of these groups to the forests, empowers their claim to the REDD+ process and to consultation and recognition of their demands. While forest-dwelling communities are not homogeneous in their views, they do frequently draw on a narrative of their intimate and long-lasting relationship with the natural environment, along with their claims of having had relative sovereignty over the forests for centuries, to demand that their privileged relationship to nature be recognized in efforts to manage forests, or to incentivize their management. Through these experiences, efforts to govern the forests through REDD+ have found sedimented resistance in the colonially racialized subjects who identify themselves along the lines of their traditional relations with the forests, and their needs for economic sustenance.

TRUTH AND DISCIPLINE IN SURINAME

During the colonial period in Suriname, indigenous and maroon communities residing in the forests were permitted by the Dutch colonial government to benefit from customary laws establishing community-led arrangements for access and use of the land. These traditional laws were recognized in peace treaties that effectively disrupted the totality of the Dutch colonial and sovereign forest governance of the territory. Hence, for some time, different swathes of forests in the territory that became Suriname were being governed through sovereign (de)colonial mentality by the colonizers and through truth (de)colonial mentality legitimized by the sovereign to the benefit of indigenous and maroon groups on the ground. These events had represented a significant shift in the internal dynamics of forest governance in Suriname during the colonial period.

The peace treaties allowing for these simultaneous governing regimes were signed between colonial rulers and indigenous peoples in the seventeenth century and maroons in the eighteenth century as confirmation of established arrangements that could be found in legal documents prohibiting settlers from molesting indigenous and maroon occupants of the land and forcing settlers to respect customary land rights.[37] The peace treaties were based on the colonizers' eventual recognition that maroon and indigenous communities had deeply rooted, culturally informed, and ongoing rights to the land.

However, this recognition was not included in the legal framework of the new Republic of Suriname, as the 1986 constitution declared that all untitled land was property of the state. Since forest communities did not hold individual titles to the land, the forest use practices of indigenous and maroon groups returned to a state of subordination under a shifting form of sovereign, though now independent, (de)colonial mentality. The independent view of community land rights that replaced the peace treaties is based on L-Decrees of 1982, which state that if there is no evidence that land within the borders of Suriname belongs to an individual, it is the domain of the state government. Access to and use of state land is evidenced by land leases that are valid for between fifteen and forty years. Actors holding these titles to state land have access only to surface resources. Resources below the soil remain under the jurisdiction of the state government.[38] While some laws in Suriname, such as the Mining Decree of 1986 and the Forestry Act of 1992, recognize the existence of communities in the forests and advocate for their livelihoods to be respected, these laws offer more of a consideration than a right. They are therefore recognized by state government bodies when convenient.

Moreover, there are no institutions that provide a venue for the concerns of indigenous or maroon communities to be voiced, as they are not recognized as legal entities within the state. This precarious situation results often in the granting of mining and logging concessions close to or overlapping with indigenous and maroon villages and the land they utilize. The circumstances of these communities then become largely dependent on the discretion of the incumbent state government of the day. Alternatively, community access to resources becomes contingent on the declarations of and adherence to tenets of social and environmental responsibility by the company to which forestry and/or mining concessions are granted.[39]

In some national legal documents, the state government of Suriname committed to resolving the land rights issue through the demarcation of indigenous and maroon lands, but this commitment was never fulfilled. After the conclusion of the civil war that took place largely in the forests of Suriname between 1986 and 1992, the Suriname government committed to resolving the claim of forest communities for land rights. But it has not done so despite having signed legally binding national-level documents and having ratified several international treaties that commit it to respecting indigenous rights.[40] Despite this, indigenous and maroon land claims remain rooted in truth (de)colonial mentality. Truth (de)colonial mentality can be seen in expressions by indigenous and maroon community members who repeatedly assert their special community relations with the forests. This truth remains as a legacy of pre-colonial forest use traditions. It harkens back to periods when indigenous and maroon land claims disrupted the sovereign colonial governance of Suriname's forest for sustained periods of time.

Policy expressions of this truth exist, although these policies often avoid essentializing indigenous and maroon communities as their direct beneficiaries. The communities are referred to instead as particularly valuable stakeholders whose capacity for engagement in REDD+ and forest conservation should be improved. Indigenous communities are often depicted as self-organized and able to legitimately represent their community interests at the national level. REDD+ implementation in Suriname also directly addresses the need to strengthen the capacity of indigenous people and maroons for engaging in the initiative. According to the country's readiness proposal, community capacity building in Suriname should therefore include securing the legitimacy of their representative organizations, consolidating the channels of communication to facilitate the sharing of information between communities and the state government and among communities themselves, training their representatives as REDD+ experts, and organizing and supporting community activities related to REDD+.[41]

Therefore, while indigenous and maroon communities are depicted in Suriname's REDD+ preparation efforts as partners to be integrated into forest governance, it is presumed that their guidance will be needed and sought in planning and other REDD+ activities.[42] This recognition demonstrates that the state government of Suriname is indeed aware of their special relationship to the forests and the land. This recognition might also signal the communities' demonstrated ability to impede state government REDD+ efforts through their outreach to international REDD+ institutions, as described in earlier chapters. In general, the relationship of forested communities with the forests is not overt in Suriname's policy documents related to REDD+, where they are instead represented as one of many social groups with which the independent sovereign state government should engage. However, the fact that they are set apart as two ethnic groups, when other ethnic groups residing in cities outside of the forests are not, indicates this special relationship.

Members of indigenous and maroon communities, however, do not hesitate to point out their relationship with the forests. They highlight that despite their continued claims for recognition of their land rights, the demands for forest conservation through REDD+ appear to take priority over the rights of forest peoples.[43] Community members express frustration at the fact that despite numerous government-led initiatives that urge them to protect the forests, the government itself keeps granting concessions that drive deforestation. In one group of communities in western Suriname, community members expressed their concerns about the Greenheart Group, a majority Chinese-owned company that is operating in the area. Communities were especially concerned about noise pollution emanating from the energy-generating machine that they fear will blow up at any time. In reference to the granting of concessions by the government, communities state that "things happen at we back, we don't even know. Papers sign and we don't even see that. Somebody will come and say, 'This is my land, this is my reservation or what you call it' and sign bam. We as representatives,[44] we can't do nothing. So, what we have to, we just only protecting where we live, the piece of land. We plant, we eat."[45] A community member in Apoera further described the experience of the indigenous people living there:

> They come in your village and they say that they are going to teach you how you must save your forests, and the government sends people to destroy your forest. We are not destroying. Yes, we must save our forest. Yes, because when you need something fresh, you go to the forest. When you need anything in the forests, you go in the forests and take it out, out of the forests. Now a next one comes in from another country and says that you must not destroy your forests, and behind our backs they come and destroy our forests for us. We must tell them that we don't want you all to come to destroy our forests, yes, but now you just come and tell me that we must not destroy our forests. We are not destroying our forests. You come and reach the forests everything and all the bush green, but now you come in Suriname, now you see the place getting very hot and this machine is coming, this coming and everything coming and destroying the forests for us.[46]

On the other hand, some other indigenous community members employed with the Greenheart company pointed to the benefits that were

Figure 13. Street view of the indigenous community of Apoera in western Suriname (Collins, 2014).

beginning to accrue to the community from the company's activities. They pointed out that the Greenheart Group was working towards donating lumber to build a market for the community and was giving local people more opportunities to gain employment.

Foreign companies operating with the permission of the state government are also causing concern to maroon communities in Nieuw Aurora, where Chinese companies are alleged to be operating in the area. According to one of my interviewees, the companies are preventing communities from accessing parts of the forests by protecting the areas with dogs. Meanwhile, communities are unsure about whether the companies and individuals even have permission to log the forests.[47] Making matters worse, a decline in economic prospects in Brazil stimulated an influx of Brazilians, who followed the gold to Suriname and set up operations outside the confines of the law. Communities in the west of Suriname are resisting this influx, explaining that they are particularly concerned about the threat to their freshwater supply and food supply from mercury that is used for mining.

Forest-dependent communities had challenged this status quo prior to the commencement of the REDD+ process in Suriname. As described in chapter 4, in 2007 the Saramaka maroons from the Upper Suriname River filed a case with the Inter-American Court of Human Rights, through which they protested violations of their traditional rights to ancestral land and won. This case was initiated by the communities after two Chinese companies destroyed homesteads and agricultural fields without offering sufficient compensation. The court decided that "the State shall delimit, demarcate, and grant collective title over the territory of the members of the Saramaka people, in accordance with their customary laws."[48] This title should eventually include rights to decide on the exploitation of natural resources such as timber and gold within that territory. In addition, the Saramaka maroons were awarded compensation from the Surinamese government for damages caused by the Chinese logging companies, to be paid into a special development fund managed by Saramaka. While the state government has made some efforts to abide by the award, it still has not legally recognized the communities as having rights to the land.[49] The inability of successive governments to satisfactorily address the issue of land rights is a bargaining chip for the communities, who draw heavily on arguments of injustice meted out to them over centuries to challenge REDD+ implementation and to demand certain political concessions to improve their overall well-being.

Suriname's land rights debate and its connection to REDD+ are most visible through the circumstances of maroon communities that were relocated to facilitate the construction of the Afobaka hydropower dam. Tellingly, it is this group of people, earlier described as now comprising the Brownsweg communities, that poses a significant threat to the forests of Suriname since some of them have turned to gold mining as their central source of income. Although they often describe the forests as their keeper and source of sustenance, stating that the forest provides for them, their idea of provision is markedly different from that of the indigenous communities, and even from some other maroon communities that were not displaced. Their idea of forest use appears to include exploitative practices with fewer concerns for the continuing well-being of the forests. Ben, the district commissioner of Brownsweg, described the situation:

> We are not waiting for money from the government. We want to decide for ourselves on what we are going to do and make our own decisions. We were moved 55 years ago. You have to see what kind of house they gave us. We had to live in it with a husband, wife, three to four children. It is too small. We want to have our own rights, make our own decisions, build our own houses, and do everything by ourselves without the government telling us what to do. . . . We try to make our own houses, so we use the rainforests. We use wood, gravel, we use everything that we can find in the rainforests, we use to build our own house.[50]

As the leader of the Gold Sector Planning Commission (OGS) mining authority in Suriname explained, once concession holders, who are usually large-scale gold miners, need the area cleared, people in the neighboring areas are relocated to different areas. The case of Koffiekamp in Suriname illustrated this, as described in chapter 2. The villagers of Koffiekamp had been initially displaced by the flooding caused by the building of the Afobaka dam and were relocated to Koffiekamp. Some years passed until the community was then again "surrounded by trucks" after the government allocated mining rights to the Canadian large-scale mining operation known as Iamgold.[51] In order to avoid further displacement, communities seek out legally recognized land rights.

One of the leaders of the maroon effort to obtain recognition for their rights over the land explained that in 2007, the Inter-American Commission on Human Rights (IACHR) recommended that free, prior, informed consent (FPIC) be obtained before any major future development on lands claimed by communities, in accordance with the Indigenous and Tribal Peoples Convention, 1989 (No. 169).[52] This maroon community member, who later joined the Surinamese state government, stated that much progress has been made. A development fund has since been created, which is jointly managed by a panel of three persons: a representative of the Saramaka, a representative of the government, and a neutral individual. He pointed to progress made on demarcating the community lands and the setting up of a bureau for land rights. He noted that the slow pace of progress was due mostly to a lack of expertise.[53] The perception within Saramaka communities, however, was that little to nothing had been done in carrying out the recommendations of the IACHR.

Indigenous groups in Suriname are also actively claiming their right to their lands. Their representative organizations, such as De Vereniging van Inheemse Dorpshoofden in Suriname (Association of Indigenous Village Leaders—VIDS), saw REDD+ implementation as a positive development because they saw REDD+ as providing an opportunity for land rights issues to be reinstated on the national agenda. They state that "having all these IGOs and NGOs focus on land rights is the biggest advantage for them [indigenous communities]."[54] VIDS therefore saw REDD+ as a means to restate their desire for land rights.

Representatives of the Ministry of Regional Development in Suriname, which works predominantly on issues related to development in forested areas, recounted the histories of these groups residing within Suriname's forests. They stated:

> Let's start with the Amerindians, the indigenous. They were living here all the time and they don't have anywhere else to go, and then during slavery, a group of Africans fought for their freedom and because they did not have weapons, they ran away and they established themselves in the forests with all kinds of consequences. Over decades, they developed their own way of life, survival system, and other kinds of systems without any help of the government at that time, you see? So, you can understand that if people have been living in a special area without the support of the government and they have been surviving all the years, until governmental influence and other things came, they feel that there is their home. Everything in the surroundings is theirs. With land rights, what they want is not to own everything but to deal with the government. If the government could deal with them and there are economic activities that can bring about a lot of development, I don't think the people will say "no we don't want that."[55]

This reflection makes the point that indigenous and maroon communities in Suriname are still reeling from the effects of the past and their perceived marginalization and are imbued with a sense of self-reliance. They remain desirous of having a voice in their own determination. For REDD+ to move forward without their resistance, these communities require recognition and progress on the issue of land rights. Drawing on perceived global injustices, some representatives of indigenous groups in Suriname, however, remained skeptical of the benefits they could accrue from REDD+. One representative stated:

> Personally, what I see is that REDD+ is a way of keeping everybody cool and big countries that have the money continue to destroy the world, so they give you a little bit of the money to say, "I give you this," but they are still doing the same thing that they used to do.[56]

These expressions of indigenous and maroon community representatives demonstrate how REDD+, as an international environmental policy, becomes embroiled in and challenged by preexisting conflict and contestation upon implementation in certain contexts. The debate over land rights is therefore not simply a REDD+ problem, but a grounded, country-specific, colonial phenomenon that is aggravated by the claims of REDD+ proponents to the carbon sequestration work of the forests. In justifying their need for land rights, communities depict the state government as being duplicitous and as a threat.

Meanwhile, state government representatives insist that REDD+ will not bring land rights. They have asserted that while the "wheelbarrows" for land rights could come through REDD+, land rights will not be granted.[57] Instead, the REDD+ procedures being developed to address grievances are posited as a possible meeting point.[58] This grievance mechanism is intended by REDD+ proponents to be a means of resolving issues related to REDD+ implementation in Suriname. These proponents also suggest that the grievance mechanism will ensure that communities are consulted before any large-scale or significant development is allowed to take place in their communities.[59] What this consultation means remains unclear, as some actors noted that it remains an open question whether these developments will go ahead if communities express concern.[60]

The challenge posed to REDD+ by the land rights issue was also expressed by representatives of the Ministry of Regional Development, who reiterated that land rights are a continuous issue for the communities, a goal for which the indigenous people and the maroons have been fighting for decades. Having this goal in mind, when REDD+ was floated as an idea, these groups saw the opportunity to restate their demands.[61] There is the frequent refrain within communities, however, that the status quo is unfair and that indigenous and forested communities were treated unjustly, not only by the former colonizers, but by the independent governments of Suriname who continue to see the natural resources the communities have conserved for centuries as available for exploitation in support of economic gain. This situation was discussed and problematized in chapter 3.

There are frequent comparisons between the perceived comfort of those living in the capital city and the lack of basic facilities in the forested or rural areas. A large section of the groups making these claims frequently refer to themselves as stewards of the forests. In so doing, they position themselves as environmentally benign forest dwellers frequently left out of the mainstream development thrust. Most often, however, these claims are unconcerned about the carbon reduction aims of REDD+.

However, truths of the special relationship that exists between forested communities and the forests have not remained static as the social dynamics within communities also change. The Organization of Indigenous People (OIP) representative explained that these communities are used to bartering as their form of exchange but that now that cash is replacing barter, there is a great need to build the capacity of the community members for managing money, along with basic needs such as school, recreation, and health facilities.[62] This shift from bartering to a greater reliance on cash has brought about a concomitant change in the way community members relate to each other. As one resident described, "We used to share fish. A big calabash of fish we would bring to your neighbor, bring to everybody, but now we sell meat to our brother, sister, father, anybody we sell."[63] When I asked why they feel the need to sell also to family members, she explained, "Everybody needs money. You don't have money, you can't buy things from the shop."[64] She went on to detail how some communities still have the culture of togetherness, such as the least modernized community in the triad of communities in western Suriname, that of Washabo, adjacent to Apoera:

> Washabo still has the tradition. They make tradition drink, culture drink. They call family around, people around and they are going to help one another to plant their farm. So, it is not her alone working on the farm. They call more people and twenty or fifteen or twelve or how much people come

Figure 14. Indigenous woman in Washabo, Suriname, making kasiri, a fermented beverage (Collins, 2014).

> . . . ten, ten ladies like, and me ah go cook to eat and maybe she buy like cigarette or whatever for the men. She bring some drink in the farm also, and everybody drink and work and thing. They plant the whole thing . . . like a community. Until Mashramami [celebration after "hard work"] when we come together. . . . So, you see everything change in our place, change.[65,66]

Now, due to the increased need for cash, indigenous communities are turning to exploitative and extractive options to gain income. These circumstances in indigenous communities are then depicted by international organizations that work in these countries as poverty that should be remedied both by directing financial and other resources towards the communities, and by shaping the residents of these communities into more suitable, market-oriented subjects who can choose rationally between income options and respond well to incentives such as those provided by REDD+. This need to govern the indigenous and forested communities in a way that makes them compatible and responsive to market concerns is by no means limited to REDD+ implementation, but the mechanism's arrival has certainly heightened this trend.

The truth that indigenous communities have special, spiritual relations with forests is also evident in Suriname.[67] Proponents of REDD+ work specifically towards integrating forest-dependent communities into the management of forests based on the recognition, explicit or otherwise, that their residence in the forests and their traditional forest use practices single these communities out as helpful for the REDD+ effort. In light of this, the role of REDD+ assistant was created, after several World Bank Forest Carbon Partnership agreement rejections of Suriname's REDD+ readiness preparation proposal, in order to discipline indigenous and maroon community members into subject positions that would make them amenable to REDD+.[68] This was brought about after some representatives of Amerindian groups challenged the REDD+ readiness proposal on the grounds that it had not been prepared in conjunction with the forested communities.

As a representative of the maroon community described:

> Thank God there were mechanisms for us to stop the R-PP from 2009 and that was approved last year . . . if we need the money, because most of them just don't look at the money, we can get like millions of dollars and then yeah, but it's not about the money, it's about the people and as long as our

> rights are not recognized, we don't know what will happen with us, we cannot agree with this part of the REDD+ process. . . . Why did we stop it? We stopped it because we were not sure that our rights were guaranteed, because the way that they prepared the R-PP, we were not happy with the language they put in it. Our main concern was the rights paragraph, the right of the people, because there was a decision that people in the interior, the maroon and the indigenous people, didn't have rights on the land. They said that our constitution was there, but we were not happy with the safeguards for our people so we couldn't agree with it.[69]

As the above quotation clearly shows, concerns about land rights formed the basis of community objections to REDD+, which led to the World Bank's rejection of Suriname's R-PP at one stage, as previously mentioned. Based on these and other objections, the Surinamese government was required by the World Bank to consult with the communities to gain their input before the R-PP was eventually approved in 2013. REDD+ assistants were subsequently identified in some communities by their chiefs and captains, as requested by the state government, and were then trained to "facilitate [*sic*] local dialogues with indigenous and maroon communities."[70] These REDD+ assistants were intended to work to "enhance the awareness and improve the collaboration and involvement in these often very remote villages."[71] Suriname's REDD+ readiness plan explained the process of integrating REDD+ assistants:

> REDD+ Assistants are . . . to be trained in conceptual understanding of REDD+. The REDD+ Assistants Collective will be used to effectively involve Indigenous and Tribal People,[72] which include soliciting the ideas and concerns of the stakeholders after they have been informed about the concept of REDD+ and the Government's plans for implementing REDD+ activities. . . . They will also be helpful to make climate change and REDD+ understandable in the local communities in between REDD+ activities.[73]

Through the creation of the role of REDD+ assistants and the associated training of forest communities to support REDD+, disciplinary (de)colonial mentality is operationalized to draw on truths associated with REDD+ governance to enroll local resource users as defenders of forests in exchange for payments. These truths were demonstrated through collectively held interpretations of communities about their codependent relationship with nature and the forests.

Figure 15. View from a highway passing through the transmigratory communities of Brownsweg (Collins, 2014).

This truth is then drawn on by REDD+ proponents as fodder for the entrenchment of market centrality for environmental management. This vision is made explicit in Suriname's R-PP, which as highlighted in chapter 4, called for the conversion of traditional knowledge into Western knowledge in support of forest cover monitoring efforts and REDD+.[74] This statement points to how the government of Suriname envisions a large-scale "seeing" of the forests from above, with forest communities imagined as complementing this effort through their experience on the ground.

In this vein, Suriname's National Plan for Forest Cover Monitoring (FCMU) depicts the role of forest dwellers similarly. The document explains how different segments of society should come together to support forest cover monitoring. While it makes clear that forest cover monitoring is a technical process that should be led by individuals versed in satellite imagery, GIS, and remote sensing, the involvement of institutions, stakeholders, and communities is described as desirable. Public participation is depicted as beneficial to the avoided-deforestation effort due to its potential to call attention to environmental problems where they occur and for improving the state governance of these issues. More specifically, the participation of forest communities in forest monitoring is presented in this document as being helpful because of its ability to increase the flow of information. Forest communities are imagined to be useful for the provision of data on nontimber forest products and the reporting of threatening activities such as gold mining in the areas near their residence. This is a particularly interesting assertion since, as previously noted, large-scale gold miners are usually in surrounding areas near residences of indigenous and maroon communities, usually with the express permission of governmental bodies, and small-scale gold miners often come from communities themselves. Through these latest efforts aimed at increasing the legibility of the forests, the people dwelling in the forests of Suriname are reimagined and disciplined through REDD+ activities.

CONCLUSION

This chapter demonstrated how the truth (de)colonial mentalities expressed by indigenous and maroon communities in Guyana and Suriname continue to manifest themselves in relation to REDD+. Primarily, these truths are manifested in the claim of forest-dependent people for access to the lands and resources that they and their ancestors have been utilizing for generations. These land claims impact REDD+ and its functioning, since the initiative depends on the allocation and establishment of clear land tenure practices. However, this clarity is elusive because some actors, who draw on some colonially maligned truths, claim that the colonial experience was an interruption of their traditional land use patterns and not a complete severance, while others claim the opposite.

Decolonization would necessitate that these truths be revisited and given priority in determining the rights to the forests of Guyana and Suriname. These truths, while often used by the now-independent governments of Guyana and Suriname to justify the special treatment and/or consideration meted out to forest-dependent communities in state governing policies, can, if taken seriously in the decolonizing endeavor, see significant segments of land (if not all) being returned to the control of indigenous communities. This would represent a significant challenge to the status quo because indigenous people can effectively lay claim to all the land that has been demarcated through the colonial enterprise as Guyana and Suriname, and perhaps even further afield. Hence, attempts to take decolonization seriously present a deep and profound challenge to the way that modern society is ordered.

However, given the scale of change that indigenous communities have historically faced, there is little likelihood that these communities and their views of themselves have not also significantly changed over time. Indigenous communities in particular have weathered European enslavement and attempts to enslave them, alliances with the colonizers that succeeded in stopping enslaved people from seeking further refuge in the forests, and even attempts to marginalize them and their truths in claiming their lands. Given these histories and experiences, a challenge can be raised to the romanticization of the past where indigeneity is automatically associated with environmental stewardship and conservation. In the words of indigenous education scholar Linda Tuhiwai Smith while looking at the image of a collection of broken eggshells, "If you didn't know the shape of an egg, how would you put those back together again? So that is the image I want you to have of what it means for me and others to talk about our [indigenous] knowledge after colonialism, that what we embarked on is this huge project of trying to put it back together again."[75] In view of the damaging effects of the colonial past and present, not only on the indigenous people, but on maroons, the natural environment, and other groups of people in these countries who were forcibly and/or coercively relocated to the Guiana Shield centuries ago, I hold back from going all the way. I advocate instead for greater recognition of how colonial histories shape land use practices in the Amazonian Guiana Shield, while using these historical injustices as a decolonial starting point.

It is important to recall here that the social dynamics of Guyana and Suriname differ from those of settler colonies, as described in the introduction to this book, where the call for recognition of the continued impact of colonialism might signal cursory or superficial alterations to a colonizer-dominant status quo. In states like Guyana and Suriname, colonized and exploited peoples vastly outnumber the direct and visible descendants of the colonizers. Therefore, a shared recognition of their colonial histories and of how society's constituent groups were pitted against each other to maintain the dominance of the white colonizer has the potential to shake the foundations of the colonial, Eurocentric order that remains and that can still be seen in modern environmental governance practices. In this fourth step towards decolonizing environmental governance in the Amazonian Guiana Shield outlined in this chapter, I advocate therefore for the prioritization of indigenous and other truths in governing efforts directed at staving off the worst effects of climate change. Indigenous claims for land must be seen as valid, as should be the claims of maroons and other groups of people disadvantaged in the formation of the independent Guyanese and Surinamese states through the colonial enterprise.

Perhaps, it is with respect for the ways in which people live and have lived with the past that we can recognize and interrogate how climate change maps onto these pasts while exacerbating their accompanying injustices. Perhaps, it is through this awareness that environmental governance policies can more accurately reflect and consider these injustices, rather than magnifying them by acting through the pretense of color/injustice blindness facilitated by reliance on markets for adjudicating amongst these concerns. Perhaps, it is through these decolonial starting points that we can find a base from which to reject the predetermined telos of development and modernization and open up space for constructing an otherwise.[76,77]

Concluding Remarks

In conclusion, I return to the question posed in the opening pages of this book: Why decolonize? I am now better placed to respond with the following answers: to open up other ways of being that do not center state governance built on oppressive and exploitative ways of relating to nature; that do not subsume human and environmental well-being within the logic of the capital-accumulating, profit-generating market; and that do not suppress non-European worldviews and environmental relations under hierarchies of race, class, gender, and other axes of difference.

This decolonizing impetus, once recognized as valuable and worthy of pursuit, should first proceed by recognizing the coloniality that continues to shape the world. This coloniality still informs dominant ways of relating to the environment, as manifested in the sovereign state in post-colonial contexts. Second, the decolonizing impetus should recognize and decenter the "store of value" ethic that was instilled in post-colonial environments during colonialism by questioning the centrality of the market as arbiter of environmental and social affairs. Third, this decolonizing force should undiscipline the racialized subject that emerged from colonial governance while taking care to consider the circumstances that have positioned certain groups to relate to nature in harmful ways.

The move towards undisciplining the subject should be holistic, rather than individualistic. It should proceed in the recognition that if environmental concerns are indeed a priority, the discipline instilled in some subjects, such as some small-scale gold miners, can be beneficial in the short term. However, a longer-term view towards undisciplining the colonial subject should be maintained towards considering the different historical interventions that have positioned certain groups over centuries in certain ways. Finally, market discipline should be countered with those collectively held truths emergent from the diverse groups of people supplanted to form patches in the fabric of multicultural Guyana and Suriname. Of course, this is only possible to the extent that these truths still exist and have not been scrubbed clean through centuries of colonialism and postcolonial continuities. To me, these are the gestures necessary for decolonizing environmental governance in the Amazonian Guiana Shield.

• • • •

In this book, I argued that global efforts to govern and to conserve tropical forests through the market, of which REDD+ is the paradigmatic example, have been stymied by under-acknowledged colonial histories and the structural conditions they engender, referred to as the coloniality.[1] During the colonial period, the forests of the Amazonian Guiana Shield functioned as places of refuge for those fleeing capital-generating activities taking place on the coasts.[2] Now, in the face of climate change, forests are again providing refuge, but this time to the international community as it attempts to stave off the worst effects of the changing climate. There is a limit to the forests' capacity for refuge for both people and environment, however. This limit is rapidly moving closer in terms of the physical sustainability of the ecosystems of the Amazon rainforests upon which forested communities and human well-being the world over depend.[3]

Since the formal close of colonialism, subsequent state governments of Guyana and Suriname have deployed policies and systems to increase the legibility of the forests and to extract their ever-changing market value. In the chapters of this book, I drew attention to the threat of climate change and to the means through which colonial histories impede and shape its address. I traced the historical process through which the market, and its accompanying ethic of extraction of value, was instilled in the places that became Guyana and Suriname. In so doing, I contextualized and outlined the efforts made by both countries to achieve development modeled on the image of the industrialized North.

In the first chapter, I suggested that REDD+ was proposed and embraced by its proponents in Guyana and Suriname because of the way that it combined promises of large influxes of cash in support of forest conservation with an ability to avoid stymieing the development ambitions of both countries. From the second chapter onwards, I looked backwards to highlight the past's continuities in the present around forest use practices through overlapping (de)colonial mentalities, as depicted in table 1. I recounted the histories that engendered the states of Guyana and Suriname, to demonstrate how people on the ground navigated their exploitative colonial pasts in ways that shape their being and current circumstances. In some cases, these histories account for the very existence of certain subjectivities, for examplc, maroons, and of the social considerations that drive their now-exploitative relationship with the forests, for example, gold mining, which is, in turn, problematized by REDD+ in its quest to avoid deforestation. Recognizing that the independent states of Guyana and Suriname are, to a large extent, expressions of coloniality, I suggest that we must behead the sovereign, that is, reduce deference to the sovereign state, if any serious attempt can be made towards recognizing other ways of being and relating to nature outside of the neoliberal, colonial world order. In this second chapter, I highlighted not only the emergence of racialized relations with the environment, but the means through which the legacy of these colonial pasts makes current forest use practices possible.

In the second chapter, I also noted that the implementation of forest-governing initiatives in this region continues to be influenced by the circumstances that shaped the historical emergence and persistence of the racialized subject. Consequently, REDD+ is not met on the ground by a subject amenable to its market-based government. Instead, REDD+ is confronted by a racialized, historically inflected subject who interacts with the environment in ways that are, in large part, shaped by colonial histories and irrational according to market logic. Still, this has not stopped proponents of REDD+ from seeking to shape societies into versions amenable to its governance. In response, subjects of REDD+ government fragment, or become increasingly incoherent in word and deed. Resistance to the instillation of market logic is most evident around the touchy issue of land rights, which goes to the heart of the colonial enterprise, and the formation of the now-independent states of Guyana and Suriname. Nevertheless, in recognizing the coloniality of these two countries and their forest governance, I note that the decolonial impetus would necessitate that the head of each sovereign state be removed from its default position as the actor through which the power to govern the forests is situated. Should the state as sovereign be beheaded, different approaches to environmental governance could become more clearly visible. The centrality of markets in the development efforts of both countries might also be more easily perceptible.

In the third chapter, I traced the processes through which the forest discursively moved, in the years following independence, further away from being a place of refuge and closer towards its current representation as a source of raw material to be exploited. This discursive transformation was not limited to the spatial confines of the forests. It was also evident on the coastlands, from where the logic of markets as arbiter of social and environmental relations was let loose. This third chapter detailed efforts to neoliberalize the wider society, along with some of the instances within which resistance to this neoliberalization can be found.

In chapter four, I identified instances where the centralization of market logic in environmental governance was resisted. There it became clear that decolonization is not all sunshine and rainbows. The trade-offs between disciplined and undisciplined subjects became evident in relation to the environment. In light of these trade-offs, it became increasingly clear that decolonization might benefit some and not others. It holds no automatic, direct, and beneficial promise for the environment. Its value for the environment and otherwise remains, however, in its potential for engaging with other ways of being that were suppressed over centuries. It gives them a chance to be taken seriously and provides much-needed alternatives to the colonially informed, modern ways of being in the world and of relating to nature that brought us to this point. After all, one thing of which one can be certain is that colonial-fueled modernity is what brought us here.[4]

In the fifth chapter, I engaged with truths, which I described as collectively held and culturally informed interpretations of the order of the world.

I argued that these truths were maligned throughout the colonial encounter. I then demonstrated their stickiness by exploring their interaction with forest governance through REDD+. Coming back to the contentious issue of land rights, I noted that taking truth (de)colonial mentalities seriously, as strategies of colonization and hence (de)colonization, might see all the land of Guyana and Suriname returning to indigenous people. This might represent a major inversion of the current status quo to the detriment of other groups that also suffered tremendously in modernity's rise.

Coloniality in Guyana and Suriname is not only indigenous land dispossession. It is the transatlantic slave trade and its legacy, racialized subjectivities, socio-environmental relations and histories, class structures, language, Eurocentrism, unquestioned Western-centric development models, economic reliance on certain commodities, market centrality in matters of collective concern, and even racism. Coloniality is also the system of policing that temporarily refused entry to the High Court in Guyana to an indigenous politician dressed in traditional garb.[5] It is the variety of Guyanese onlookers of that incident who deemed his traditional attire inappropriate and vulgar and commented to this effect publicly on social media platforms.

Hence, while some might fairly expect calls for decolonization to automatically and directly lead to the conclusion that lands must be returned to indigenous people, I point here to the fact that different groups of people were also involuntarily relocated to the lands that became Guyana and Suriname. These people were exploited in different ways and to different degrees throughout colonial enterprise. Colonization was the coerced relocation and racialization of people *alongside* the claiming of land. Surely, an argument that hinges on the awareness that indigenous people are not, and do not have to be, assessed along an externally imposed metric of intactness should be open to the fact that indigeneity is not a fixed, racial grouping that automatically makes its members more capable of environmental stewardship than others. Instead, what does tend to privilege indigeneity in discussions of environmental governance is the proximity of indigenous communities to nature and their demonstrated histories of living with and in the forests in largely sustainable ways. This way of life is, however, not the exclusive preserve of indigenous communities racialized as such.

Table 2 Multiple and Overlapping (De)colonial Mentalities in Amazonian Guiana Shield

Period	*Dominant (De)colonial Mentality*	*Characteristics*
Pre-Colonial Period (before 1600s)		Traditional land use patterns
Colonial Period (early 1600s to late 1900s)	Sovereign	Colonial government in Guyana and Suriname
	Truth	Constantly present and shifting modes of relating to nature in indigenous communities, intermittently institutionalized in Suriname
Post-Independence Forest Use (late 1900s to present)	Sovereign	Colonial government becoming less-coercive independent governments in Guyana and Suriname
	Truth	Forest communities challenging independent states of Guyana and Suriname
	Disciplinary	Market logic being spread to resistant or overlooked pockets in society, especially in Guyana
REDD+ (2009 to present)	Sovereign	Independent governments becoming primary REDD+ actors in Guyana and Suriname
	Truth	Forest communities challenging REDD+ as continued incursion into their forests in both countries
	Disciplinary	REDD+ being made palatable to forest communities through REDD+ assistants in Suriname and community development officers in Guyana
	Neoliberal	REDD+ being employed to incentivize behavior change in Guyana and Suriname

This observation therefore challenges the apparent commonsense logic of some onlookers that indigenous people, racialized as such, are better able to protect nature than others from the outset. In other words, if we remove the shackles of race and essentialisms around the concept of indigeneity, we must also make some room for the disassociation of that racial construct from positions that see it as being inherently more sustainable than others. When this disassociation is combined with an awareness that European colonization disadvantaged a variety of groups outside of those indigenous to the Amazonian Guiana Shield, and relegated others to having to sustain themselves in the forests, decolonization could be understood as entailing more than an automatic return of land to those communities residing there before the arrival of European colonizers. Decolonization could instead be expanded to hold space for ways of being that do not default to capitalist and colonially charted pathways.

Hence, I hold back from going all the way down this "commonsense" route and advocate instead that the seriousness of indigenous and other non-European truths be accepted, rather than continuously relegated to a state of subservience and backwardness to be remedied. I remind the astute reader that a politics of recognizing nonsettler colonial histories and their deeply entrenched relationship with the environment, the state, and forest governance strategies in Guyana and Suriname differs markedly from the politics of settler colonial states, such as those of the United States and Australia. In Guyana and Suriname, the colonizer has simply gone home, mostly leaving the groups they oppressed to fight for control over the ruins that remain. In settler states, the fight for land between settlers and indigenous groups is more immediate, direct, and pressing.

In developing these arguments throughout the book, I questioned the feasibility of market-based tools for encouraging conservation in the postcolonial, developing world. I argued that the marketization logic that supports neoliberal conservation runs counter to the space needed to conceptualize ways of being outside of the overarching neoliberal coloniality of today.[6] Other ways of being are crowded out by the logics of the market that are increasingly being relied upon in the dominant approaches to conserving nature in the face of climate change.[7] I advocated, too, for more nuance in the literature that addresses processes of racialization within critiques of the Anthropocene, by arguing that the category of blackness, positioned in opposition to whiteness, subsumes and overlooks the different ways in which people often categorized as black are differentially exposed to and affected by both climate change and efforts to govern and combat it.

Finally, this project demonstrated that reflections on how conservation and environmental governance initiatives are colonized in implementation also provide space for showing how decolonization can be pursued on the ground.[8] This latter potential can be envisioned through Foucauldian thought that allows us to pinpoint the continuities in governing strategies over time. It is important to remember, though, that governmental interventions are often unsuccessful, in that their outcomes are not always those that were originally intended by the governing body. Decolonization should therefore not solely be pursued as an outcome in itself. Instead, it should be pursued as a means of tackling established, dominant, and persistent colonial governing strategies and of departing from the paths they are known to have set out—although this departure will surely have unintended consequences of its own. In other words, decolonization should be pursued as continual movement away from the colonial present towards an unspecified end point, rather than as a path that can take us in direct and logical steps from point A to B.

While I maintain that decolonization should be pursued as an explicit recognition of and movement away from colonialism's continued structuring of global power distribution patterns, and of environmental relations and policies today, I acknowledge that in today's global power constellation, my propositions to behead the sovereign, to decenter markets, to undiscipline subjects, and to counter discipline with truths are difficult to accept *simultaneously*. Thus, as a bare minimum, I suggest that conservation actors and those imbued with the power to govern forests in post-colonial environments take consistent steps towards avoiding the repetition and perpetuation of those forest-governing strategies characteristic of colonial claims on land and people that I marked out in this book.

Nonetheless, while we wait for this decolonizing impetus to pick up speed, I recall here that Guyana and Suriname have recently found significant amounts of oil in their territorial waters. Effectively, the economic priorities of their state governments have moved from forest conservation in the name of climate change, but covertly in the name of economic development, to oil extraction. These priorities have skipped the stage of asserting oil's magnifying effects on climate change but continue to operate in the pursuit of economic development. These discoveries, made well after the close of my data-gathering period, have profound implications for REDD+ as a tool for meeting the development expectations that some may have had for it.[9] Governments are not choosing between REDD+ and extractive activity, but are putting in place conditions that allow REDD+ and extraction to take place side by side.

The intensification of extraction in this region further demonstrates that REDD+ continues to be pursued even in places where carbon-emitting activities, namely the production of oil, are deepening their hold. Development, on the other hand, continues to be pursued in ways that are modeled on the image of the West. If environmental concerns are to become a priority, an inversion of that order is necessary. Decolonization presents a profound challenge to this order. It urges reflection and demands a departure from racialized, capitalist-infused modernity as an end goal in and of itself. Decolonization demands a bold, audacious reckoning with the injustices of the past in order to make space for other ways of being in the present.

Notes

INTRODUCTION

1. Heemskerk, *Rights to Land and Resources for Indigenous Peoples and Maroons in Suriname*; Sizer and Price, *Backs to the Wall in Suriname.*

2. Price, "Uneasy Neighbors."

3. Bovolo et al., "The Guiana Shield Rainforests"; van Kuijk, *REDD+ Development in the Guianas.*

4. Bovolo et al., "The Guiana Shield Rainforests."

5. Relative to other forested countries around the world.

6. Hook, "Following REDD+."

7. Angelsen, "REDD+ as Result-Based Aid."

8. Angelsen, "REDD+ as Result-Based Aid."

9. Fletcher et al., "Questioning REDD+ and the Future of Market-Based Conservation."

10. Büscher and Fletcher, *The Conservation Revolution.*

11. Fanon, *The Wretched of the Earth*, 36.

12. Collins, "Colonial Residue."

13. See United Nations, Climate Change, "What Is REDD+?"

14. Phelps, Friess, and Webb, "Win–Win REDD+ Approaches Belie Carbon–Biodiversity Trade-Offs."

15. Office of the President, Republic of Guyana, "Creating Incentives to Avoid Deforestation."

16. Greenpeace, *Bad Influence—How McKinsey-Inspired Plans Lead to Rainforest Destruction.*

17. Peluso and Vandergeest, "Genealogies of the Political Forest and Customary Rights in Indonesia, Malaysia, and Thailand"; Vandergeest and Peluso, "Political Forests."

18. Rodney, *A History of the Guyanese Working People.*

19. Mintz, *Sweetness and Power.*

20. Mintz, *Sweetness and Power.*

21. Knight, *The Caribbean;* Moore, "Sugar and the Expansion of the Early Modern World-Economy."

22. Glasgow, *Guyana.*

23. Richardson and Eltis, *Atlas of the Transatlantic Slave Trade.*

24. Patterson, *Slavery and Social Death.*

25. Robinson, *Black Marxism.*

26. Price, "Uneasy Neighbors."

27. Price, "Uneasy Neighbors."

28. Price, "Uneasy Neighbors."

29. Some research indicates that maroon groups had developed in Guyana but had since been wiped out (Thompson, *Flight to Freedom*). More informal reports indicate that communities of runaway slaves continue to exist in Guyana but have not been officially recognized as such (intergovernmental organization representative, interviewed in 2014).

30. Like Guyana, Suriname was colonized by both the Dutch and the British at different intervals.

31. This term refers to the intention of the government to move the communities to more permanent domiciles in the future, a move that never took place.

32. REDD+ is divided into three phases: readiness, implementation, and payments for results. When I use the term *implementation,* however, I am referring to the entire process of enacting REDD+ in these places.

33. Gebara and Agrawal, "Beyond Rewards and Punishments in the Brazilian Amazon"; McGregor et al., "Beyond Carbon, More than Forest?"; Boer, "Welfare Environmentality and REDD+ Incentives in Indonesia."

34. Haraway, "Anthropocene, Capitalocene, Plantationocene, Chthulucene"; Ferdinand, *Decolonial Ecology.*

35. Escobar, *Encountering Development;* Goldstein, "Decolonialising 'Actually Existing Neoliberalism.'"

36. Tuck and Yang, "Decolonization Is Not a Metaphor."

37. Tuck and Yang, "Decolonization Is Not a Metaphor."

38. The difference between settler and nonsettler colonialism is, like much else, a question of degree. Make no mistake, some amount of settling is involved in nonsettler colonialism, but the underpinning rationale of nonsettler colonialism is not to settle overseas. Instead, nonsettler forms of colonization aim to take

control of land and resources and to co-opt their use for the benefit of the mother country.

39. Williams, *Capitalism and Slavery.*

40. It is important to note, however, that Foucault argues that governing strategies scarcely ever bring about their desired outcomes. Therefore, (de)colonization understood through strategies of governance in the Foucauldian sense should be pursued in the knowledge that the end goal of decolonization could and should be imagined and worked towards but remains unpredictable in terms of its potential for achieving a particular, linear outcome.

41. Parreñas, *Decolonizing Extinction;* Anthias, *Limits to Decolonization;* Whyte, "Settler Colonialism, Ecology, and Environmental Injustice."

42. Quijano, "Coloniality of Power and Eurocentrism in Latin America."

43. Post-colonial theory also addresses these continuities. It shows that colonialism was co-constitutive of both the colonizer and the colonized and demonstrates the ways in which the colonial experience continues to shape the present. Post-colonialism was a reflective critical endeavor that sought out colonialism's continued presence in the work of national elites post-independence. It also sought to identify the colonial effect on ideas of nationalism, knowledge, democracy, and other areas of society in the now independent states. Decolonial theory, originating in Latin America, also recognizes these continuities but with a view to dismantling them and their resultant power structures. See Chari and Verdery, "Thinking between the Posts"; Shohat, "Notes on the 'Post-Colonial'"; Grossberg, "On Postmodernism and Articulation"; Rattansi, "Postcolonialism and Its Discontents"; Prakash, "Writing Post-Orientalist Histories of the Third World"; Simon, "Separated by Common Ground?"

44. Quijano, "Coloniality of Power and Eurocentrism in Latin America."

45. Quijano, "Coloniality of Power and Eurocentrism in Latin America"; Maldonado-Torres, "On the Coloniality of Being"; Wynter, "Unsettling the Coloniality of Being/Power/Truth/Freedom."

46. See Hall, "The West and the Rest" in Gupta et al., *Race and Racialization.*

47. Mignolo and Walsh, *On Decoloniality,* 109.

48. Knight, *The Caribbean.*

49. I hasten to note, though, that my intent in representing interconnectivity in this way is not to make Guyana and Suriname the "center" of capitalism and the world's affairs, but to illustrate the reverberating ramifications of their histories as they spread around the world.

50. Wolfe, *Traces of History.*

51. Quijano, "Coloniality of Power and Eurocentrism in Latin America."

52. Williams, *Capitalism and Slavery*; Robinson, *Black Marxism.*

53. Mignolo and Walsh, *On Decoloniality;* Quijano, "Coloniality of Power and Eurocentrism in Latin America."

54. Quijano, "Coloniality of Power and Eurocentrism in Latin America."

55. Mollett, "The Power to Plunder."

56. Quijano, "Coloniality of Power and Eurocentrism in Latin America."

57. Quijano, "Coloniality of Power and Eurocentrism in Latin America."

58. Capitalized as in Quijano, "Coloniality of Power and Eurocentrism in Latin America."

59. Quijano, "Coloniality of Power and Eurocentrism in Latin America," 217–18.

60. Quijano, "Coloniality of Power and Eurocentrism in Latin America," 217–18.

61. Saldanha, "Race," 453.

62. Wolfe, *Traces of History.*

63. Wolfe, *Traces of History.*

64. Wolfe, *Traces of History.*

65. Wolfe, *Traces of History.*

66. Wolfe, *Traces of History.*

67. See, for exception, Yusoff, *A Billion Black Anthropocenes or None.*

68. Sundberg, "Placing Race in Environmental Justice Research in Latin America."

69. Sundberg, "Placing Race in Environmental Justice Research in Latin America."

70. Johnson, *Becoming Creole.*

71. See Bureau of Statistics Guyana, *Guyana Population and Housing Census: Preliminary Report 2012*; Household Budget Survey, *Suriname Census 2012.*

72. Knight, *The Caribbean.*

73. In *Decolonial Ecology,* Ferdinand uses *colonial inhabitation* to refer to the mode of inhabiting the Earth that was violently implemented through European colonization of the Americas.

74. Mignolo and Walsh, *On Decoloniality.*

75. Federici, *Caliban and the Witch.*

76. Malm, *Fossil Capital.*

77. Ferdinand, *Decolonial Ecology.*

78. Mignolo and Walsh, *On Decoloniality.*

79. Quijano, "Coloniality of Power and Eurocentrism in Latin America."

80. Anti-colonial acts took place during the colonial period.

81. Decolonial acts seek to dismantle the structural legacy of colonialism.

82. Büscher and Fletcher, *The Conservation Revolution;* Brockington, Duffy, and Igoe, *Nature Unbound;* Büscher, Dressler, and Fletcher, *Nature Inc.*

83. Büscher and Fletcher, *The Conservation Revolution;* Brockington, Duffy, and Igoe, *Nature Unbound.*

84. Collins et al., "Plotting the Coloniality of Conservation."

85. Adams and Mulligan, *Decolonizing Nature;* Asiyanbi, "A Political Ecology of REDD+"; Domínguez and Luoma, "Decolonising Conservation Policy"; Neumann, *Imposing Wilderness;* Fairhead and Leach, *Reframing Deforestation;* Kashwan et al., "From Racialized Neocolonial Global Conservation to an Inclusive and Regenerative Conservation."

86. Büscher and Fletcher, *The Conservation Revolution.*

87. I recognize that using a European thinker to investigate decolonization might be a provocative move to some. However, I see decolonial thinking, and post-colonial thinking for that matter, as rejecting the overrepresentation of European knowledge systems in matters related to the Global South, or as delinking from assumptions of its universality. This does not necessitate a complete rejection of European knowledge systems on the whole.

88. Collins, "How REDD+ Governs."

89. Collins, "Guyana in the Eye of the Storm in 2021."

90. While some tensions exist between Marxist-influenced decolonial theory (by way of decolonial theory's inheritance from world systems theory) and the poststructural thinking of Foucault, I see these two schools of thought as complementary for my purposes. The *post* in *poststructural* thinking serves as a recognition of the structural concerns of Marxists and signals a movement past these concerns to "explain how colonial power worked through forms of difference that were produced in part through colonial encounters and through representations of 'self' and 'other,' with lasting practical effects" (Chari and Verdery, "Thinking between the Posts," 25). Hence, the structural outcome of the colonial experience observed by decolonial thinkers can productively be parsed with poststructural thinking, as I attempt to do here. For a thorough substantiation of the complementarity of this approach in conceptualizations of power, see Castro-Gómez, Kopsick, and Golding, "Michel Foucault and the Coloniality of Power." There, Gómez argues that decolonial research should not be held hostage to macrostructural thought and interpretations of power that depend exclusively on the thorough dismantling of capitalism for freedom from racism and other colonially rooted societal ills to be achieved. Gómez writes that "to contend that the only valid strategy of decolonization centres on 'abolishing capitalism' through revolution betrays an ignorance of decolonial struggles at multiple levels of generality. . . . Therefore, Foucault's greatest contribution to decolonial theory is perhaps the argument that molar analyses, although necessary, run the risk of arriving at a sort of methodological platonism that favours secular trends and longue durée changes while overlooking micro-agencies at the level of body and affect" (Castro-Gómez, Kopsick, and Golding, "Michel Foucault and the Coloniality of Power," 14).

91. Foucault, *The Birth of Biopolitics.*

92. See Agrawal, *Environmentality;* Fletcher, "Neoliberal Environmentality"; Miller and Rose, "Governing Economic Life."

93. Dean, *Governmentality.*

94. Foucault, *The Birth of Biopolitics.*

95. Okereke, Bulkeley, and Schroeder, "Conceptualizing Climate Governance beyond the International Regime," 71.

96. Castree, "Neoliberalism and the Biophysical Environment."

97. Fletcher, "Neoliberal Environmentality."

98. Fletcher, "Neoliberal Environmentality," 173.

99. Li, *Fixing Non-Market Subjects.*

100. Fletcher, "Neoliberal Environmentality."

101. Fletcher and Breitling, "Market Mechanism or Subsidy in Disguise?"

102. Fletcher, "Neoliberal Environmentality."

103. In the European sense of the word.

104. Foucault, *The Birth of Biopolitics,* 311.

105. Quijano, "Coloniality of Power and Eurocentrism in Latin America."

106. While this might seem out of step with more recent British norms, the reader could be reminded, in the words of Cedric Robinson, that in the colonies "the sexual and moral customs of the Black lower classes, for all their vitality and attractiveness, amounted to a rejection of English bourgeois sensibility, they were an affront to the morality of thc colonial model set before the natives" (Robinson, *Black Marxism,* 252).

107. Collins, "Colonial Residue."

108. Foucault, *Power/Knowledge.*

109. Büscher and Fletcher, *The Conservation Revolution.*

110. News Source Guyana, "Exxon Announces Its 15th Oil Discovery in Guyana."

111. Offshore Energy Today, "Apache Cheers 'Significant Oil Discovery' Offshore Suriname."

112. The race against environmental disaster I refer to here is not mere hyperbole. Deforestation has only increased since 2014, although more consistently in Suriname. However, some allowance should be made for improvements in monitoring deforestation rates (see globalforestwatch.org).

CHAPTER 1

1. Malm, *Fossil Capital,* 8.

2. Harris, *World Ethics and Climate Change.*

3. Pachauri and Reisinger, *IPCC Fourth Assessment Report.*

4. Stern, *The Economics of Climate Change.*

5. Bumpus and Liverman, "Accumulation by Decarbonization and the Governance of Carbon Offsets."

6. Agarwal and Narain, *Global Warming in an Unequal World.*

7. Malm, *Fossil Capital*, 1.
8. IPCC, 2007. *Climate Change 2007.*
9. Pachauri and Reisinger, *IPCC Fourth Assessment Report.*
10. Caribbean Community Climate Change Centre, *Climate Change and the Caribbean.*
11. Pachauri and Reisinger, *IPCC Fourth Assessment Report.*
12. Pachauri and Reisinger, *IPCC Fourth Assessment Report*, 28.
13. Caribbean Community Climate Change Centre, *Climate Change and the Caribbean.*
14. Ellis, "Suriname and the Chinese."
15. Environmental NGO representative, interviewed in 2014.
16. State government advisor, interviewed in 2014.
17. Small-scale gold miners from Brazil are referred to as "garimpieros."
18. "Pork knocker" is another term used to refer to small-scale gold miners in Guyana.
19. Leading figure in gold and diamond mining in Guyana, interviewed in 2014.
20. Leading figure in gold and diamond mining in Guyana, interviewed in 2014.
21. Hook, "Following REDD+."
22. This is because both countries are highly forested, boasting forest covers of over 85%.
23. Green, "U.S. Providing Aid to Victims of Heavy Flooding in Suriname."
24. Graham, "Flooding Slams Suriname."
25. Green, "U.S. Providing Aid to Victims of Heavy Flooding in Suriname."
26. Caribbean News Now, "Caribbean Net News: Disease Raises Death Toll to 33 in Flood-Stricken Guyana."
27. Pan American Health Organization, "Floods in Guyana—January/February 2005."
28. Office of the President, Republic of Guyana, "Creating Incentives to Avoid Deforestation," 1.
29. Office of the President, Republic of Guyana, "Creating Incentives to Avoid Deforestation," 1.
30. Office of the President, Republic of Guyana, "Creating Incentives to Avoid Deforestation," 1.
31. Office of the President, Republic of Guyana, "Creating Incentives to Avoid Deforestation," 1.
32. IPCC, 2007. *Climate Change 2007.*
33. Cenamo et al., *Casebook of REDD Projects in Latin America.*
34. Cenamo et al., *Casebook of REDD Projects in Latin America.*
35. Cenamo et al., *Casebook of REDD Projects in Latin America.*
36. Wertz-Kanounnikoff et al., "How Can We Monitor, Report and Verify Carbon Emissions from Forests?"

37. AIPP and IWGIA, *Briefing Paper on REDD+, Rights and Indigenous Peoples.*

38. Parker et al., *The Little REDD Book.*

39. AIPP and IWGIA, *Briefing Paper on REDD+, Rights and Indigenous Peoples.*

40. Bradley, "Does Community Forestry Provide a Suitable Platform for REDD?"

41. Naughton-Treves and Day, *Lessons about Land Tenure, Forest Governance and REDD+.*

42. Forsyth, *Multilevel, Multiactor Governance in REDD+.*

43. Angelsen et al., "Learning from REDD+"; Fletcher et al., "Questioning REDD+ and the Future of Market-Based Conservation"; Fletcher et al., "Debating REDD+ and Its Implications."

44. Angelsen, "REDD+ as Result-Based Aid"; Angelsen et al., "Learning from REDD+."

45. Dramatic in terms of the development of the bilateral agreement with Norway marketed as a pilot project and the receipt of numerous awards for championing the initiative received by the country's leader.

46. WWF, "FCPF Approves Suriname's REDD+ Readiness Preparation Proposal."

47. *Suriname Readiness Preparation Proposal.*

48. Brockington, Duffy, and Igoe, *Nature Unbound.*

49. Though Büscher and Fletcher argue that in practice conservation arose in tandem with capitalism's spread, being a component of rather than challenge to, its dominance in their book: Büscher and Fletcher, *The Conservation Revolution.*

50. Büscher, Dressler, and Fletcher, *Nature Inc.*

51. Büscher, Dressler, and Fletcher, *Nature Inc.*, 3.

52. Fletcher and Büscher, "The PES Conceit."

53. Harvey, *NeoLiberalism.*

54. West and Brockington, "Introduction."

55. Büscher, Dressler, and Fletcher, *Nature Inc.*, 4.

56. Castree, "Neoliberalism and the Biophysical Environment."

57. Larner, "Neo-Liberalism."

58. Mudge, "What Is Neo-Liberalism?'

59. Castree sees neoliberalism as being interpreted differently within its philosophic, program, and policy domains, though divisions exist even within these realms. Useful in Castree's work is the recognition that neoliberalism is differentiated in its manifestations around the globe and that its spread is uneven, existing in hybrid forms. He uses the term *neoliberalism* to reflect that neoliberalism is an ongoing process related to three things: "first, a worldview (i.e., a body of normative principles, goals, and aspirations amounting to a philosophy of life, or

something close to one); second, a policy discourse (i.e., a set of specific values, norms, ambitions, and associated policy proposals professed by those who control, or realistically seek to control, the formal apparatuses of government); and, third, a set of policy measures (i.e., concrete regulations and procedures that make both the worldview and the policy discourse evident in some tangible way)" (Castree, "Neoliberalism and the Biophysical Environment").

60. Larner, "Neoliberalism?"; Read, "A Genealogy of Homo-Economicus."

61. Van Hecken et al., "Silencing Agency in Payments for Ecosystem Services (PES) by Essentializing a Neoliberal 'Monster' into Being."

62. Harvey, "Neo-Liberalism as Creative Destruction."

63. Starting from the conceptualization of these policies as market based, Fletcher and Büscher follow the literature that charts their implementation on the ground, while recognizing that their implementation would never, in fact, align with their conceptualization. Other researchers, however, starting from the implementation of the policies on the ground, note their inconsistency with global interpretations of market-based policy and use this inconsistency to challenge the conceptualization of PES as market based in the first place. In response, while Fletcher and Büscher accept the inconsistency in execution, they caution against connecting this inconsistency to an argument that PES is not market based in nature and caution that unclarified assumptions about the nature of neoliberalism lie at the center of this debate. They write that "even if on-the-ground implementation of PES often lacks substantial neoliberal mechanics the approach should still be considered an important element of a global program to spread neoliberalism as a particular rationality and mode of capital accumulation" (Fletcher and Büscher, "The PES Conceit," 224).

In response, Van Hecken et al., proceeding from grounded analyses of PES in implementation, attribute the inconsistencies between its implementation and its label as neoliberal conservation, or market-based instruments, as sufficient for challenging its identification as such. In a paper that misconstrued the arguments made by Fletcher and Büscher in "The PES Conceit" to a significant extent, the authors responded by arguing that the approach of Fletcher and Büscher was "obscuring the complexity and situational history, practice and scale of the processes involved" (Van Hecken et al., "Silencing Agency in Payments for Ecosystem Services," 314). Their misconstruction was demonstrated by their (mis)use of the term *neoliberal governmentality*, their attribution to governmentality with concerns for failure and success, their attribution of "what emerges from PES interventions" as "neoliberal conceit" (Van Hecken et al., "Silencing Agency," 314), and their uncritical representation of neoliberalism as purely structural and hegemonic in ways that demonstrate a lack of awareness of its nuances (see Castree, "Neoliberalism and the Biophysical Environment," for example).

While this debate stems, in part, from disciplinary perspectives and the degree to which one prioritizes grounded and lived realities over policy formulation within institutional power structures, it proceeded in a way that restricted potential for constructive debate on the ample meeting ground between the two approaches embodied by how to interpret the significant difference between vision and execution in environmental governance policies (Carrier and West, *Virtualism, Governance and Practice*). Fletcher and Büscher emphasize the connecting thread and logic of PES with a particular vision of environmental governance, while Van Hecken et al. use execution to challenge the validity of interpretations of that vision. The arguments in this book proceed from the interpretation of neoliberalism and, as a corollary, neoliberal conservation, as not solely hegemonic, proceeding more in line with the varied aspects of neoliberalism outlined by Castree and its governing logics outlined by Foucault.

64. Fletcher, "How I Learned to Stop Worrying and Love the Market."
65. Doane, "From Community Conservation to the Lone (Forest) Ranger."
66. Fletcher and Breitling, "Market Mechanism or Subsidy in Disguise?"
67. Bigger et al., "Reflecting on Neoliberal Natures."
68. For an exception, see Collins, "Colonial Residue."
69. Collins et al., "Plotting the Coloniality of Conservation," 15.
70. Collins et al., "Plotting the Coloniality of Conservation," 15.
71. Collins et al., "Plotting the Coloniality of Conservation," 16.
72. Trafford, "Empire's New Clothes," 39.
73. Though this is not the case between Guyana and Norway.
74. Parreñas, *Decolonizing Extinction*, 7.

CHAPTER 2

1. Guyana Ministry of Finance. *National Development Strategy*, 27.
2. Svensson, *National Plan for Forest Cover Monitoring*.
3. Hofman et al., "Stage of Encounters," 590.
4. Emmer, "Caribbean Plantations and Indentured Labour, 1640–1917."
5. Emmer, "Caribbean Plantations and Indentured Labour, 1640–1917."
6. Williams, *Capitalism and Slavery*.
7. Foucault, *The Birth of Biopolitics*.
8. Glasgow, *Guyana*.
9. Griffiths and La Rose, *Searching for Justice and Land Security*.
10. Heemskerk, *Demarcation of Indigenous and Maroon Lands in Suriname*.
11. Colchester, *Guyana*.
12. Colchester, *Guyana*.
13. Reporter, "Amerindian Tribes of Guyana."
14. Emmer, "Caribbean Plantations and Indentured Labour, 1640–1917."
15. Federici, *Caliban and the Witch*.

16. Handler and Reilly, "Contesting 'White Slavery' in the Caribbean."
17. Handler and Reilly, "Contesting 'White Slavery' in the Caribbean," 35.
18. Handler and Reilly, "Contesting 'White Slavery' in the Caribbean."
19. Dowlah, *Cross-Border Labor Mobility,* 119.
20. Dowlah, *Cross-Border Labor Mobility.*
21. Emmer, "Caribbean Plantations and Indentured Labour, 1640–1917."
22. Patterson, *Slavery and Social Death,* 159.
23. Dowlah, *Cross-Border Labor Mobility.*
24. Colchester, *Guyana.*
25. Josiah, *Migration, Mining, and the African Diaspora.*
26. *Bush* is a colloquial word for forests.
27. Colchester, *Guyana.*
28. Colchester, *Guyana.*
29. Colchester, *Guyana.*
30. Ross, as cited in Dabydeen and Samaroo, *Across the Dark Waters,* 29.
31. This term has also been used to refer to Chinese indentured servants (Williams, *Capitalism and Slavery*) and was loosely used over time to refer to laborers regardless of origin.
32. Colchester, *Guyana.*
33. Colchester, *Guyana.*
34. By "modern" here, I am pointing to the fact that the coasts were seen to be more similar to the colonial metropole than to the forests.
35. Mistry et al., *Up-Scaling Support for Community Owned Solutions.*
36. Glasgow, *Guyana.*
37. Rodney, *A History of the Guyanese Working People.*
38. Josiah, *Migration, Mining, and the African Diaspora.*
39. Josiah, *Migration, Mining, and the African Diaspora.*
40. Singh, "Re-Democratization in Guyana and Suriname."
41. Menke and Egger, *Country Report Suriname.*
42. Rabe, *US Intervention in British Guiana.*
43. Colchester, *Guyana.*
44. Colchester, *Guyana.*
45. Vaughn, "Reconstructing the Citizen," 363–64.
46. Rabe, *US Intervention in British Guiana.*
47. Leading figure in gold and diamond mining in Guyana, interviewed in 2014.
48. Daly, *The Making of Guyana.*
49. Jagan, *The West on Trial.*
50. Dominant figure in regulating the mining industry, interviewed in 2014.
51. Dominant figure in regulating the mining industry, interviewed in 2014.
52. Intergovernmental organization representative, interviewed in 2014.
53. Lowenthal, "The Range and Variation of Caribbean Societies."

54. Lowenthal, "The Range and Variation of Caribbean Societies."
55. Lowenthal, "The Range and Variation of Caribbean Societies," 53.
56. Moore, "Sugar and the Expansion of the Early Modern World-Economy."
57. Colchester, *Guyana.*
58. Colchester, *Guyana.*
59. Jagan, *The West on Trial.*
60. Colchester, *Guyana.*
61. Haden, *Forestry Issues in the Guiana Shield Region.*
62. The Private Sector Commission of Guyana, Conservation International-Guyana, and WWF (Guianas), "Bridging Guyana's Mining Sector with Its Low Carbon Development Path."
63. The Private Sector Commission of Guyana, Conservation International-Guyana, and WWF (Guianas), "Bridging Guyana's Mining Sector with Its Low Carbon Development Path."
64. The Private Sector Commission of Guyana, Conservation International-Guyana, and WWF (Guianas), "Bridging Guyana's Mining Sector with Its Low Carbon Development Path."
65. Laws of Guyana, Forests Act, 7.
66. Guyana Forestry Commission, *Guyana National Forest Plan.*
67. Guyana Forestry Commission, *Guyana National Forest Plan.*
68. Menke and Egger, *Country Report Suriname;* Singh, "Re-Democratization in Guyana and Suriname."
69. Government advisor on climate change, interviewed in 2014.
70. At the time of the interview.
71. Government advisor on climate change, interviewed in 2014.
72. Similar to the knowledge of indigenous people in Guyana.
73. Colchester, *Forest Politics in Suriname;* Heemskerk, *Demarcation of Indigenous and Maroon Lands in Suriname.*
74. Colchester, *Forest Politics in Suriname.*
75. Kambel and MacKay, *The Rights of Indigenous Peoples and Maroons in Suriname.*
76. Fatah-Black, *White Lies and Black Markets.*
77. Struiken and Healy, "Suriname."
78. Colchester, *Forest Politics in Suriname.*
79. Fatah-Black, *White Lies and Black Markets.*
80. Hoefte, *Suriname in the Long Twentieth Century.*
81. Choenni, "Ethnicity and Politics."
82. Colchester, *Forest Politics in Suriname.*
83. Hoogbergen and Kruijt, "Gold, 'Garimpeiros' and Maroons."
84. Mhango, "The Political Economy of Aid."
85. Mhango, "The Political Economy of Aid."
86. Mhango, "The Political Economy of Aid."

87. Apoera is one of the field sites I visited while collecting data for my research.

88. Mhango, "The Political Economy of Aid."

89. Mhango, "The Political Economy of Aid."

90. Ronnie Brunswijk is, at the time of writing in 2023, the current vice president of Suriname.

91. Price, "The Maroon Population Explosion."

92. Garimpeiros are small-scale, Brazilian gold miners.

93. Hoogbergen and Kruijt, "Gold, 'Garimpeiros' and Maroons."

94. Hoogbergen and Kruijt, "Gold, 'Garimpeiros' and Maroons."

95. Surinamese maroon community member, interviewed in 2014.

96. Ben, a Surinamese maroon community member, interviewed in 2014.

97. Ben, a Surinamese maroon community member, interviewed in 2014.

98. LMR, an indigenous community member, interviewed in 2014.

99. Ben, a Surinamese maroon community member, interviewed in 2014.

100. Again, I am not commenting on whether indigenous or maroon communities are more or less culturally intact. In fact, my entire argument hinges on the fact that people are positioned in certain ways in relation to colonial histories and demands for labor, both historically and in the present. My argument depends on avoiding essentialisms of different groups of people. Rather, it shows how people negotiate certain power dynamics to take on identities and to cement practices in view of their environments.

101. I point out here that my intention is not to use these quotations to support an argument for or against the relative "intactness" of the cultures of these two sets of forest communities.

102. de Bruijne and Schalkwijk, "The Position and Residential Patterns of Ethnic Groups in Paramaribo's Development in the Twentieth Century," 240.

103. Colchester, *Forest Politics in Suriname.*

104. Colchester, *Forest Politics in Suriname.*

105. Colchester, *Forest Politics in Suriname.*

106. Kambel and MacKay, *The Rights of Indigenous Peoples and Maroons in Suriname.*

107. Hoogbergen and Kruijt, "Gold, 'Garimpeiros' and Maroons."

108. Kambel and MacKay, *The Rights of Indigenous Peoples and Maroons in Suriname.*

109. Hoogbergen and Kruijt, "Gold, 'Garimpeiros' and Maroons."

110. World Bank, "Suriname Overview."

111. Bruijne and Schalkwijk, "The Position and Residential Patterns of Ethnic Groups in Paramaribo's Development in the Twentieth Century," 253.

112. Bruijne and Schalkwijk, "The Position and Residential Patterns of Ethnic Groups in Paramaribo's Development in the Twentieth Century," 253.

113. Academic respondent in Suriname, interviewed in 2014.

114. Hoogbergen and Kruijt, "Gold, 'Garimpeiros' and Maroons."
115. Ellis, "Suriname and the Chinese."
116. Ellis, "Suriname and the Chinese."
117. Ellis, "Suriname and the Chinese."
118. John, a Surinamese creole person, in a casual remark during a 2014 interview.
119. Svensson, *National Plan for Forest Cover Monitoring.*
120. Svensson, *National Plan for Forest Cover Monitoring.*
121. Ministry of Labour, Technological Development and Environment, *Suriname Second National Communication to the United Nations Framework Convention on Climate Change.*
122. Heemskerk, *Rights to Land and Resources for Indigenous Peoples and Maroons in Suriname.*
123. Heemskerk, *Rights to Land and Resources for Indigenous Peoples and Maroons in Suriname.*
124. Menke and Egger, *Country Report Suriname.*
125. Menke and Egger, *Country Report Suriname.*
126. Collins, "Racing Climate Change in Guyana and Suriname."
127. Crutzen and Stoermer, "Global Change Newsletter."
128. Büscher and Fletcher, *The Conservation Revolution.*
129. Luke, "Tracing Race, Ethnicity, and Civilization in the Anthropocene."
130. Moore, *Capitalism in the Web of Life.*
131. Yusoff, *A Billion Black Anthropocenes or None.*
132. Luke, "Tracing Race, Ethnicity, and Civilization in the Anthropocene"; Yusoff, *A Billion Black Anthropocenes or None.*
133. Yusoff, *A Billion Black Anthropocenes or None.*
134. Haraway, *Staying with the Trouble.*
135. Büscher and Fletcher, *The Conservation Revolution.*
136. They refer to "mainstream conservation," but I use just "conservation" here instead.
137. Büscher and Fletcher, *The Conservation Revolution,* 4.
138. Büscher and Fletcher, *The Conservation Revolution,* 72.
139. Büscher and Fletcher, *The Conservation Revolution,* 116.
140. Erickson, "Anthropocene Futures."
141. I reiterate here that my aim is not to refute the focus on whiteness in critical approaches to interrogating the Anthropocene and in directing calls for responsibility, but rather to draw attention to racialized differences within the too-simple category of "blackness."
142. Collins, "Racing Climate Change in Guyana and Suriname."
143. Rodney, *A History of the Guyanese Working People.*
144. Rodney, *A History of the Guyanese Working People,* 2.
145. Rodney, *A History of the Guyanese Working People,* 3–4.

146. Quijano, "Coloniality of Power and Eurocentrism in Latin America."

147. Jones, "Contesting the Boundaries of Gender, Race and Sexuality in Barbadian Plantation Society."

148. Collins, "Racing Climate Change in Guyana and Suriname."

149. Wolfe, *Traces of History.*

150. This is perhaps out of necessity given their global scale and temporal span.

151. Peluso and Vandergeest, "Genealogies of the Political Forest and Customary Rights in Indonesia, Malaysia, and Thailand."

152. I use "limitations" here instead of "dependencies" to demonstrate my view that the path can still be changed.

CHAPTER 3

1. WWF, *Community-Based Monitoring.*

2. Bernstein, "Modernization Theory and the Sociological Study of Development."

3. Rodney, *How Europe Underdeveloped Africa;* Mignolo and Walsh, *On Decoloniality;* Quijano, "Coloniality of Power and Eurocentrism in Latin America."

4. Quijano, "Coloniality of Power and Eurocentrism in Latin America."

5. Scott, *Seeing Like a State.*

6. Scott, *Seeing Like a State.*

7. Brighenti, "Visibility a Category for the Social Sciences."

8. Brighenti, "Visibility a Category for the Social Sciences," 326.

9. Brighenti, "Visibility a Category for the Social Sciences," 327.

10. Brighenti, "Visibility a Category for the Social Sciences," 329.

11. Brighenti, "Visibility a Category for the Social Sciences," 326.

12. Brighenti, "Visibility a Category for the Social Sciences."

13. Svensson, *National Plan for Forest Cover Monitoring,* 10.

14. Svensson, *National Plan for Forest Cover Monitoring.*

15. Barquin et al., *The Knowledge and Skills Needed to Engage in REDD+,* 94.

16. Barquin et al., *The Knowledge and Skills Needed to Engage in REDD+.*

17. Barquin et al., *The Knowledge and Skills Needed to Engage in REDD+.*

18. Das and Poole, *Anthropology in the Margins.*

19. Asad, "Ethnographic Representation, Statistics and Modern Power."

20. Barquin et al., *The Knowledge and Skills Needed to Engage in REDD+.*

21. Scott, *Seeing Like a State.*

22. Ferguson, *The Anti-Politics Machine.*

23. Norad, *Real-Time Evaluation of Norway's International Climate and Forest Initiative Synthesising Report 2007–2013.*

24. Dooley and Griffiths, *Indigenous Peoples' Rights, Forests and Climate Policies in Guyana*, 15.

25. Guyana Lands and Surveys Commission, *Guyana National Land Use Plan.*

26. Guyana Lands and Surveys Commission, *Guyana National Land Use Plan.*

27. GGMC representative interviewed in 2014.

28. Office of the President, Republic of Guyana, "Low Carbon Development Strategy Update."

29. Office of the President, Republic of Guyana, "Low Carbon Development Strategy Update."

30. Office of the President, Republic of Guyana, "Low Carbon Development Strategy Update."

31. Indufor, and Guyana Forestry Commission, *Guyana REDD+ Monitoring Reporting & Verification System.*

32. Best, *International Conference on the Impact of REDD+ for HFLD Countries.*

33. Sarika, a representative of the GFC, interviewed in 2014.

34. Sarika, a representative of the GFC, interviewed in 2014.

35. Svensson, *National Plan for Forest Cover Monitoring.*

36. Svensson, *National Plan for Forest Cover Monitoring.*

37. Das and Poole, *Anthropology in the Margins.*

38. SN, who worked with the Guyana government on its environmental strategy, interviewed in 2014.

39. Office of the President, Republic of Guyana, "Creating Incentives to Avoid Deforestation."

40. GGMC representative interviewed in 2014.

41. Arets et al., *Towards a Carbon Balance for Forests in Suriname.*

42. WWF, "FCPF Approves Suriname's REDD+ Readiness Preparation Proposal."

43. Arets et al., *Towards a Carbon Balance for Forests in Suriname.*

44. Arets et al., *Towards a Carbon Balance for Forests in Suriname*, 9.

45. Arets et al., *Towards a Carbon Balance for Forests in Suriname.*

46. Amy, a GFC official interviewed in 2014.

47. GB, respondent in a 2014 interview.

48. See Rodney, *A History of the Guyanese Working People, 1881–1905.*

49. Escobar, *Encountering Development.*

50. McKinsey & Company, "Who We Are."

51. Greenpeace, *Bad Influence.*

52. Office of the President, Republic of Guyana, "Creating Incentives to Avoid Deforestation," 2.

53. *Suriname Readiness Preparation Proposal.*

54. Collins, "How REDD+ Governs."

55. Griffiths and Anselmo, *Indigenous Peoples and Sustainable Livelihoods in Guyana.*

56. Griffiths and Anselmo, *Indigenous Peoples and Sustainable Livelihoods in Guyana.*

57. Environmental NGO representative 2, interviewed in 2014.

58. Joe, a representative of an indigenous group implementing REDD+ forest conservation activities, interviewed in 2014.

59. Tour guide, interviewed in 2014.

60. Crew member in Surama, interviewed in 2014.

61. Dominant figure in regulating the mining industry, interviewed in 2014.

62. Larry, an indigenous community leader, interviewed in 2014.

63. Larry, an indigenous community leader, interviewed in 2014.

64. District commissioner, interviewed in 2014.

65. International conservation organization director, interviewed in 2014.

66. International conservation organization director, interviewed in 2014.

67. Collins, "How REDD+ Governs."

68. Van Hecken et al., "Silencing Agency in Payments for Ecosystem Services."

69. International conservation organization director, interviewed in 2014.

70. Sarika, a representative of the GFC, interviewed in 2014.

71. Sarika, a representative of the GFC, interviewed in 2014.

72. Doane, "From Community Conservation to the Lone (Forest) Ranger."

73. OIS, a representative of a conservation organization, interviewed in 2014.

74. International conservation organization director, interviewed in 2014.

75. Chapin, "A Challenge to Conservationists."

76. Chapin, "A Challenge to Conservationists."

77. Angelsen, "REDD+ as Result-Based Aid."

78. Büscher, "Anti-Politics as Political Strategy."

79. Brockington, "Ecosystem Services and Fictitious Commodities."

80. Castree, "Neoliberalism and the Biophysical Environment."

81. Forest Carbon Partnership Facility, *Guyana Readiness Preparation Proposal.*

82. Forest Carbon Partnership Facility, *Guyana Readiness Preparation Proposal*, 9.

83. Forest Carbon Partnership Facility, *Guyana Readiness Preparation Proposal*, 9.

84. Forest Carbon Partnership Facility, *Guyana Readiness Preparation Proposal*, 9.

85. Collins, "How REDD+ Governs."

86. Government of Guyana, *Enhancing National Competitiveness.*

87. Scott, *Seeing Like a State.*

88. Government of Guyana, *Enhancing National Competitiveness*, 6.

89. Intergovernmental organization representative, interviewed in 2014.

90. Vezzoli, *The Evolution of Surinamese Emigration across and beyond Independence*.

91. Hout, "Development under Patrimonial Conditions"; Academic respondent in Suriname, interviewed in 2014.

92. Government of Guyana, *Enhancing National Competitiveness*, 14, emphasis added.

93. Government of Guyana, *Enhancing National Competitiveness*.

94. The Private Sector Commission of Guyana, Conservation International-Guyana, and WWF (Guianas), "Bridging Guyana's Mining Sector with Its Low Carbon Development Path."

95. Government of Guyana, *Enhancing National Competitiveness*, 17.

96. Bank of Guyana, *Annual Report*.

97. Bank of Guyana, *Annual Report*.

98. Centrale Bank van Suriname, *Jaarverslag 2014*.

99. International Bank for Reconstruction and Development, *Suriname Sector Competitive Analysis*.

100. Collins, "How REDD+ Governs."

101. Brockington, "Ecosystem Services and Fictitious Commodities."

102. *Undisciplined* in the Foucauldian sense used here means "not disciplined" and is not intended to be a qualitative judgment along an axis of good or bad.

CHAPTER 4

1. *Suriname Readiness Preparation Proposal*.

2. Foucault, *The Birth of Biopolitics*.

3. Fletcher, "Neoliberal Environmentality."

4. What I am pointing to here is the process through which characteristics attributed to certain "races" are projected onto people, resulting in subjects who are racialized in different ways to create an abstract racialized subject, rather than successfully racialized people. In other words, through pluralization, the characteristics being attributed to the individual subject are made more solid. Whereas the racialized subject in its singularity, and in its use to refer to several groups of historically supplanted people, acknowledges that people who were brought to the Guiana Shield were stripped of their prior ethnic affiliations and identities and christened anew, so to speak, in ways peculiar to colonialism's capital-accumulating logic. Hence, in maintaining the singularity of the racialized individual and collective subject, I maintain an emphasis on shared physical and social displacement through the colonial experience.

5. Saldanha, "Race."

6. Myers et al., "Messiness of Forest Governance"; McGregor et al., "Beyond Carbon, More than Forest?"

7. Collins, "Colonial Residue."

8. Foucault, *The Birth of Biopolitics.*

9. These assumptions are entirely hypothetical.

10. Li, "Governmentality."

11. Li, "Governmentality."

12. JJ, a UNDP employee in the Suriname office, interviewed in 2014.

13. JJ, a UNDP employee in the Suriname office, interviewed in 2014.

14. Li originally uses the term *governmentality,* but for my purposes, I recast it as (de)colonial mentality to refer to both the strategies of colonization and decolonization, as previously outlined.

15. TIGY, a representative of a civil society organization in Guyana, interviewed in 2014.

16. TIGY, a representative of a civil society organization in Guyana, interviewed in 2014.

17. TIGY, a representative of a civil society organization in Guyana, interviewed in 2014.

18. Tropenbos International representative, interviewed in 2014.

19. Tropenbos International representative, interviewed in 2014.

20. Tropenbos International representative, interviewed in 2014.

21. Intergovernmental organization representative, interviewed in 2014.

22. Dooley and Griffiths, *Indigenous Peoples' Rights, Forests and Climate Policies in Guyana.*

23. Dooley and Griffiths, *Indigenous Peoples' Rights, Forests and Climate Policies in Guyana,* 86.

24. Dooley and Griffiths, *Indigenous Peoples' Rights, Forests and Climate Policies in Guyana.*

25. Anita, an academic in Suriname, interviewed in 2014.

26. Anita, an academic in Suriname, interviewed in 2014.

27. Anita, an academic in Suriname, interviewed in 2014.

28. Anita, an academic in Suriname, interviewed in 2014.

29. Dalby, "The Geopolitics of Climate Change."

30. Dalby, "The Geopolitics of Climate Change," 42.

31. Intergovernmental organization representative, interviewed in 2014.

32. Intergovernmental organization representative, interviewed in 2014.

33. Representative of the Ministry of Regional Development, interviewed in 2014.

34. Paul, interviewed in 2014.

35. Office of the President, Republic of Guyana, "Low Carbon Development Strategy Update."

36. Leading figure in gold and diamond mining in Guyana, interviewed in 2014.

37. Leading figure in gold and diamond mining in Guyana, interviewed in 2014.

38. GGMC officers, interviewed in 2014.

39. Miner, interviewed in 2014.

40. George, a miner in St. Luce, interviewed in 2014.

41. Lemke, "'The Birth of Bio-Politics': Michel Foucault's Lecture."

42. George, a miner in St. Luce, interviewed in 2014.

43. George, a miner in St. Luce, interviewed in 2014.

44. Transparency International (Guyana) representative, interviewed in 2014.

45. Rick, representative of the Gold Sector Planning Commission (OGS) mining authority in Suriname, interviewed in 2014.

46. Rick, representative of the Gold Sector Planning Commission (OGS) mining authority in Suriname, interviewed in 2014.

47. Rick, representative of the Gold Sector Planning Commission (OGS) mining authority in Suriname, interviewed in 2014.

48. Li, "Governmentality."

49. Woodworker, interviewed in 2014; a community member, interviewed in 2014.

50. Stearman, "Revisiting the Myth of the Ecologically Noble Savage in Amazonia."

51. WR, a maroon member of an organization representing the interests of miners in Suriname, interviewed in 2014.

52. George, a small-scale miner, interviewed in 2014.

53. George, a small-scale miner, interviewed in 2014.

54. Peterson and Heemskerk, "Deforestation and Forest Regeneration Following Small-Scale Gold Mining in the Amazon."

55. Peterson and Heemskerk, "Deforestation and Forest Regeneration Following Small-Scale Gold Mining in the Amazon."

56. Government advisor on climate change, interviewed in 2014.

57. Leading figure in gold and diamond mining in Guyana, interviewed in 2014.

58. Although this situation is arguably changing on account of Guyana's new status as an oil producer.

59. Representative of a wood company operating in the area, interviewed in 2014.

60. Mistry et al., *Up-Scaling Support for Community Owned Solutions;* WWF, *Community-Based Monitoring, Reporting and Verification Know-How.*

61. Mistry et al., *Up-Scaling Support for Community Owned Solutions.*

62. Captain Dorps, edited 2014 interview.

63. Rick, representative of the Gold Sector Planning Commission (OGS) mining authority in Suriname, interviewed in 2014.

64. Rick, representative of the Gold Sector Planning Commission (OGS) mining authority in Suriname, interviewed in 2014.

65. WR, a maroon member of an organization representing the interests of miners in Suriname, interviewed in 2014.

CHAPTER 5

1. Representative of a maroon community in Suriname, at an HFLD conference in 2014.

2. Representative of a maroon community in Suriname, at an HFLD conference in 2014.

3. Foucault, *The Birth of Biopolitics*.

4. Robinson, *Black Marxism*.

5. Federici, *Caliban and the Witch*.

6. Doane, "From Community Conservation to the Lone (Forest) Ranger."

7. Office of the President, Guyana, "A Low-Carbon Development Strategy."

8. GRIF, *United Nations Development Programme Country: GUYANA Project Document*.

9. GRIF, *United Nations Development Programme Country: GUYANA*.

10. Li, *Fixing Non-Market Subjects*.

11. Li, *Fixing Non-Market Subjects*.

12. GRIF, *United Nations Development Programme Country: GUYANA*, 3.

13. GRIF, *United Nations Development Programme Country: GUYANA*, 3.

14. GRIF, *United Nations Development Programme Country: GUYANA*.

15. Intergovernmental organization representative, interviewed in 2014.

16. "Community Development Plan—Hobodeia—Region # 1."

17. Skutsch and Turnhout, "How REDD+ Is Performing Communities."

18. GRIF, *United Nations Development Programme Country: GUYANA*, 3.

19. GRIF, *United Nations Development Programme Country: GUYANA*, 4.

20. GRIF, *United Nations Development Programme Country: GUYANA*, 5.

21. See Hulme, *Global Poverty* for an explanation of different measures of poverty.

22. Ferguson, *The Anti-Politics Machine*.

23. Dooley and Griffiths, *Indigenous Peoples' Rights, Forests and Climate Policies in Guyana*.

24. APA representatives, interviewed in 2014.

25. APA representatives, interviewed in 2014.

26. APA representatives, interviewed in 2014.

27. APA representatives, interviewed in 2014.

28. Government of Guyana. *A Low Carbon Development Strategy: Transforming Guyana's Economy While Combating Climate Change.*

29. Dooley and Griffiths, *Indigenous Peoples' Rights, Forests and Climate Policies in Guyana.*

30. Dooley and Griffiths, *Indigenous Peoples' Rights, Forests and Climate Policies in Guyana*; Hook, "Mapping Contention."

31. Guyana Lands and Surveys Commission, *Guyana National Land Use Plan.*

32. Guyana Lands and Surveys Commission, *Guyana National Land Use Plan.*

33. Dooley and Griffiths, *Indigenous Peoples' Rights, Forests and Climate Policies in Guyana.*

34. Leading figure in gold and diamond mining in Guyana, interviewed in 2014.

35. Guyana Lands and Surveys Commission, *Guyana National Land Use Plan.*

36. Office of the President, Republic of Guyana, "Low Carbon Development Strategy Update."

37. Heemskerk, *Rights to Land and Resources for Indigenous Peoples and Maroons in Suriname*, 3.

38. Heemskerk, *Rights to Land and Resources for Indigenous Peoples and Maroons in Suriname.*

39. Heemskerk, *Rights to Land and Resources for Indigenous Peoples and Maroons in Suriname.*

40. Heemskerk, *Rights to Land and Resources for Indigenous Peoples and Maroons in Suriname.*

41. *Suriname Readiness Preparation Proposal.*

42. GRIF, *United Nations Development Programme Country: GUYANA.*

43. Maroon representative, interviewed in 2014.

44. The speaker's position was altered to hide their identity.

45. Community representatives of Apoera, interviewed in 2014.

46. Community member in Apoera, interviewed in 2014.

47. Linus, a maroon community member, interviewed in 2014.

48. Heemskerk, *Demarcation of Indigenous and Maroon Lands in Suriname*, 14.

49. Heemskerk, *Demarcation of Indigenous and Maroon Lands in Suriname*, 14.

50. Ben, district commissioner of Brownsweg, interviewed in 2014.

51. Rick, representative of the Gold Sector Planning Commission (OGS) mining authority in Suriname, interviewed in 2014.

52. Linus, a maroon community member, interviewed in 2014.

53. Linus, a maroon community member, interviewed in 2014.

54. VIDS, interviewed in 2014.
55. Ministry of Regional Development in Suriname, interviewed in 2014.
56. OIS, an indigenous group representative, interviewed in 2014.
57. Kurt, a state government representative, interviewed in 2014.
58. Kurt, a state government representative, interviewed in 2014.
59. Kurt, a state government representative, interviewed in 2014.
60. State government representative, interviewed in 2014.
61. Ministry of Regional Development, interviewed in 2014.
62. OIP representative, interviewed in 2014.
63. Community member in Washabo, interviewed in 2014.
64. Community member in Washabo, interviewed in 2014.
65. Repetition is being used here for emphasis.
66. Community member in Washabo, interviewed in 2014.
67. McGregor et al., "Beyond Carbon, More than Forest?"
68. OIS, an indigenous group representative, interviewed in 2014.
69. Linus, a representative of the maroon community, interviewed in 2014.
70. Best, *International Conference on the Impact of REDD+ for HFLD Countries,* 24.
71. Svensson, *National Plan for Forest Cover Monitoring,* 53.
72. I note here that within Suriname's policy documents, the term *tribal* is often used to refer to maroon communities. Like many other labels referring to different groups of people, both terms are subject to claims that they are pejorative, often by those who study these areas from a North American perspective. Meanwhile, both terms are used in the official policy documents in Suriname, the area in this study that features recognized "tribal" or "maroon" groups.
73. *Suriname Readiness Preparation Proposal,* 20–21.
74. *Suriname Readiness Preparation Proposal.*
75. Tuhiwai Smith, "Decolonising Methodologies, 20 Years On."
76. Parreñas, *Decolonizing Extinction.*
77. See Buscher and Fletcher, *The Conservation Revolution* for an example of other ways of being.

CONCLUDING REMARKS

1. Quijano, "Coloniality of Power and Eurocentrism in Latin America."
2. This function was one of many important ecological functions that should not be overlooked.
3. Bovolo et al., "The Guiana Shield Rainforests."
4. Mignolo and Walsh, *On Decoloniality;* Maldonado-Torres, "On the Coloniality of Being"; Malm, *Fossil Capital;* Haraway, "Anthropocene, Capitalocene, Plantationocene, Chthulucene."

5. Stabroek News, "Shuman Initially Prevented from Attending High Court in Indigenous Attire."

6. Fletcher, *Failing Forward.*

7. Buscher and Fletcher, *The Conservation Revolution.*

8. Requena-i-Mora and Brockington, "Seeing Environmental Injustices."

9. Collins, "The Extractive Embrace."

References

"A Low Carbon Development Strategy: Transforming Guyana's Economy While Combating Climate Change. Draft Report, Region 8, Kato." Consultation Report, 2009. http://www.lcds.gov.gy/index.php/documents/reports/local/consultations-reports/186-sub-national-consultation-draft-report-region-8-kato/file.

Adams, William (Bill), and Martin Mulligan. *Decolonizing Nature: Strategies for Conservation in a Post-Colonial Era.* London: Routledge, 2002. https://doi.org/10.4324/9781849770927.

Agarwal, Anil, and Sunita Narain. *Global Warming in an Unequal World: A Case of Environmental Colonialism.* New Delhi: Centre for Science and Environment, 1991. https://cdn.cseindia.org/userfiles/GlobalWarming%20Book.pdf.

Agrawal, Arun. *Environmentality: Technologies of Government and the Making of Subjects.* Durham, NC: Duke University Press, 2005.

AIPP and IWGIA. *Briefing Paper on REDD+, Rights and Indigenous Peoples: Lessons from REDD+ Initiatives in Asia.* Asian Indigenous Peoples Pact (AIPP) and International Work Group for Indigenous Affairs (IWGIA), 2012. https://www.iwgia.org/images/publications/0655_Doha_briefing_Final_Artork.pdf.

Angelsen, Arild. "REDD+ as Result-Based Aid: General Lessons and Bilateral Agreements of Norway." *Review of Development Economics* 21, no. 2 (2017): 237–64. https://doi.org/10.1111/rode.12271.

Angelsen, Arild, Maria Brockhaus, Amy E. Duchelle, Anne Larson, Christopher Martius, William D. Sunderlin, Louis Verchot, Grace Wong, and Sven Wunder. "Learning from REDD+: A Response to Fletcher et al." *Conservation Biology* 31, no. 3 (2017): 718–20. https://doi.org/10.1111/cobi.12933.

Anthias, Penelope. *Limits to Decolonization: Indigeneity, Territory, and Hydrocarbon Politics in the Bolivian Chaco.* Ithaca, NY: Cornell University Press, 2018.

Arets, E. J. M. M., B. Kruijt, K. Tjon, V. P. Atmopawiro, R. F. Kanten, S. Crabbe, O. S. Bánki, and S. Ruysschaert. *Towards a Carbon Balance for Forests in Suriname.* Wageningen, Netherlands: Alterra, 2011. http://library.wur.nl/WebQuery/wurpubs/421156.

Asad, Talal. "Ethnographic Representation, Statistics and Modern Power." *Social Research* 61, no. 1 (Spring 1994): 55–88.

Asiyanbi, Adeniyi P. "A Political Ecology of REDD+: Property Rights, Militarised Protectionism, and Carbonised Exclusion in Cross River." *Geoforum* 77 (2016): 146–56.

Bank of Guyana. *Annual Report.* 2015. https://www.bankofguyana.org.gy/bog/images/research/Reports/ANNREP2015.pdf.

Barquin, L., M. Chacón, S. N. Panfil, A. Adeleke, E. Florian, and R. Triraganon. *The Knowledge and Skills Needed to Engage in REDD+: A Competencies Framework.* Arlington, VA: Conservation International, Centro Agronómico Tropical de Investigación y Enseñanza, International Union for the Conservation of Nature, Regional Community Forestry Training Center, 2014.

Bernstein, Henry. "Modernization Theory and the Sociological Study of Development." *The Journal of Development Studies* 7, no. 2 (1971): 141–60.

Best, Lisa. *International Conference on the Impact of REDD+ for HFLD Countries.* Meeting Report, 2014. https://www.surinameredd.org/media/1150/hfld-conference-suriname.pdf.

Bigger, Patrick, Jessica Dempsey, Adeniyi Asiyanbi, Kelly Kay, Rebecca Lave, Becky Mansfield, Tracey Osborne, Morgan Robertson, and Gregory Simon. "Reflecting on Neoliberal Natures: An Exchange: The Ins and Outs of Neoliberal Natures." *Environment and Planning E: Nature and Space* 1, no. 1–2 (2018).

Boer, Henry. "Welfare Environmentality and REDD+ Incentives in Indonesia." *Journal of Environmental Policy & Planning* 19 (2017): 1–15.

Bovolo, C. Isabella, Thomas Wagner, Geoff Parkin, David Hein-Griggs, Ryan Pereira, and Richard Jones. "The Guiana Shield Rainforests—Overlooked Guardians of South American Climate." *Environmental Research Letters* 13, no. 7 (2018): 074029. https://doi.org/10.1088/1748-9326/aacf60.

Bradley, Amanda. "Does Community Forestry Provide a Suitable Platform for REDD? A Case Study from Oddar Meanchey, Cambodia." *Lessons about Land Tenure, Forest Governance and REDD+. Case Studies from*

Africa, Asia and Latin America, 61–72. USAID, 2012. www.rmportal.net/landtenureforestsworkshop.

Brighenti, Andrea. "Visibility a Category for the Social Sciences." *Current Sociology* 55, no. 3 (2007): 323–42.

Brockington, Dan. "Ecosystem Services and Fictitious Commodities." *Environmental Conservation* 38, no. 4 (December 2011): 367–69. https://doi.org/10.1017/S0376892911000531.

Brockington, Dan, Rosaleen Duffy, and Jim Igoe. *Nature Unbound: Conservation, Capitalism and the Future of Protected Areas.* Oxford, UK: Earthscan, 2008.

Bruijne, Ad, and Aart Schalkwijk. "The Position and Residential Patterns of Ethnic Groups in Paramaribo's Development in the Twentieth Century." *NWIG: New West Indian Guide / Nieuwe West-Indische Gids* 79, no. 3/4 (2005): 239–71.

Bumpus, Adam G., and Diana M. Liverman. "Accumulation by Decarbonization and the Governance of Carbon Offsets." *Economic Geography* 84, no. 2 (2008): 127–55.

Bureau of Statistics Guyana. *Guyana Population and Housing Census: Preliminary Report 2012.* 2012. https://www.statisticsguyana.gov.gy/download.php?file=88.

Büscher, Bram. "Anti-Politics as Political Strategy: Neoliberalism and Transfrontier Conservation in Southern Africa." *Development and Change* 41, no. 1 (2010): 29–51.

Büscher, Bram, Wolfram Dressler, and Robert Fletcher. *Nature Inc.: Environmental Conservation in the Neoliberal Age.* Tucson: University of Arizona Press, 2014.

Büscher, Bram, and Robert Fletcher. *The Conservation Revolution: Radical Ideas for Saving Nature beyond the Anthropocene.* London: Verso Trade, 2020.

Caribbean Community Climate Change Centre. *Climate Change and the Caribbean: A Regional Framework for Achieving Development Resilient to Climate Change (2009–2015).* [https://www.preventionweb.net/publication/climate-change-and-caribbean-regional-framework-achieving-development-resilient-climate]: PreventionWeb, July 2009.

Caribbean News Now. "Caribbean Net News: Disease Raises Death Toll to 33 in Flood-Stricken Guyana." Caribbean News Now, 2005. http://www.caribbeannewsnow.com/caribnet/2005/02/11/death.shtml.

Carrier, J. G., and P. West (Eds.). *Virtualism, Governance and Practice: Vision and Execution in Environmental Conservation,* volume 13. Oxford, NY: Berghahn Books, 2009.

Castree, Noel. "Neoliberalism and the Biophysical Environment: A Synthesis and Evaluation of the Research." *Environment and Society: Advances in Research* 1, no. 1 (2010): 5–45.

Castro-Gómez, Santiago, Kyle Kopsick, and David Golding. "Michel Foucault and the Coloniality of Power." *Cultural Studies* 37, no. 3 (2023): 444–60. https://doi.org/10.1080/09502386.2021.2004435.

Cenamo, M. C., M. NOGUEIRA Pavan, M. T. Campos, Ana Cristina Barros, and Fernanda Carvalho. *Casebook of REDD Projects in Latin America*. Manaus, Brazil: IDESAM and The Nature Conservancy, 2009.

Centrale Bank van Suriname. *Jaarverslag 2014*. 2014. https://www.cbvs.sr/.

Chapin, Mac. "A Challenge to Conservationists." *World Watch* (November /December 2004): 17–31.

Chari, Sharad, and Katherine Verdery. "Thinking between the Posts: Postcolonialism, Postsocialism, and Ethnography after the Cold War." *Comparative Studies in Society and History* 51, no. 01 (January 2009): 6–34. https://doi.org/10.1017/S0010417509000024.

Choenni, Chan E. S. "Ethnicity and Politics: Political Adaption of Hindostanis in Suriname." *Sociological Bulletin* 63, no. 3 (2014): 407–31.

Colchester, Marcus. *Forest Politics in Suriname*. Utrecht: International Books, 1996.

———. *Guyana: Fragile Frontier*. Kingston, Jamaica: Latin America Bureau, 1997.

Collins, Yolanda Ariadne. "Colonial Residue: REDD+, Territorialisation and the Racialized Subject in Guyana and Suriname." *Geoforum* 106 (1 November 2019): 38–47. https://doi.org/10.1016/j.geoforum.2019.07.019.

———. "Guyana in the Eye of the Storm in 2021: Convergence, Stasis and Reverberation." *Revista de Ciencia Política (Santiago)* 42, no. 2 (2022): 333–54. https://doi.org/10.4067/s0718-090x2022005000113.

———. "How REDD+ Governs: Multiple Forest Environmentalities in Guyana and Suriname." *Environment and Planning E: Nature and Space* 3, no. 2 (2020): 323–45. https://doi.org/10.1177/2514848619860748.

———. "Racing Climate Change in Guyana and Suriname." *Politics* 43, no. 2 (2021): 186–200. https://doi.org/10.1177/02633957211042478.

———. "The Extractive Embrace: Shifting Expectations of Conservation and Extraction in the Guiana Shield." *Environmental Politics* 31, no. 8 (2021): 1–23.

Collins, Yolanda Ariadne, Victoria Maguire-Rajpaul, Judith E. Krauss, Adeniyi Asiyanbi, A. Jiminez, Mathew Bukhi Mabele, and Mya Alexander-Owen. "Plotting the Coloniality of Conservation." *Journal of Political Ecology* 28, no. 1 (2021).

"Community Development Plan—Hobodeia—Region # 1," 2011.

Crutzen, Paul J., and Eugene F. Stoermer. "Global Change Newsletter." *The Anthropocene* 41 (2000): 17–18.

Dabydeen, David, and Brinsley Samaroo. *Across the Dark Waters: Ethnicity and Indian Identity in the Caribbean*. London: Macmillan Caribbean, 1996.

Dalby, Simon. "The Geopolitics of Climate Change." *Political Geography* 37 (2013): 38–47.

Daly, Vere T. *The Making of Guyana*. London: Macmillan, 1974.

Das, Veena, and Deborah Poole. *Anthropology in the Margins*. Santa Fe, NM: SAR Press, 2004. https://muse.jhu.edu/book/24179.

de Bruijne, A., and A. Schalkwijk. "The Position and Residential Patterns of Ethnic Groups in Paramaribo's Development in the Twentieth Century." *New West Indian Guide / Nieuwe West-Indische Gids, 79*, no. 3–4 (2005): 239–71.

Dean, Mitchell. *Governmentality: Power and Rule in Modern Society*. Thousand Oaks, CA: SAGE, 2009.

Doane, Molly. "From Community Conservation to the Lone (Forest) Ranger: Accumulation by Conservation in a Mexican Forest." *Conservation and Society* 12, no. 3 (2014): 233–44.

Domínguez, Lara, and Colin Luoma. "Decolonising Conservation Policy: How Colonial Land and Conservation Ideologies Persist and Perpetuate Indigenous Injustices at the Expense of the Environment." *Land* 9, no. 3 (2020): 65.

Dooley, Kate, and Tom Griffiths. *Indigenous Peoples' Rights, Forests and Climate Policies in Guyana—A Special Report*. Georgetown, Guyana: Amerindian Peoples Association, 2014.

Dowlah, Caf. *Cross-Border Labor Mobility: Historical and Contemporary Perspectives*. Cham, Switzerland: Palgrave Macmillan, 2020.

Ellis, R. Evan. "Suriname and the Chinese: Timber, Migration, and Less-Told Stories of Globalization." *SAIS Review of International Affairs* 32, no. 2 (2012): 85–97.

Emmer, Pieter C. "Caribbean Plantations and Indentured Labour, 1640–1917: A Constructive or Destructive Deviation from the Free Labour Market?" *Itinerario* 21, no. 1 (1997): 73–97.

Erickson, Bruce. "Anthropocene Futures: Linking Colonialism and Environmentalism in an Age of Crisis." *Environment and Planning D: Society and Space* 38, no. 1 (2018): 111–28. https://doi.org/10.1177/0263775818806514.

Escobar, Arturo. *Encountering Development: The Making and Unmaking of the Third World* (Student ed.). Princeton, NJ: Princeton University Press, 1995. http://www.jstor.org/stable/j.ctt7rtgw.

Fairhead, James, and Melissa Leach. *Reframing Deforestation: Global Analyses and Local Realities: Studies in West Africa* (1st ed.). London: Routledge, 1998.

Fanon, Frantz. *The Wretched of the Earth*. Translated from the French by Constance Farrington. New York: Grove Press, 1963.

Fatah-Black, Karwan. *White Lies and Black Markets: Evading Metropolitan Authority in Colonial Suriname, 1650–1800*. Boston: Brill, 2015.

Federici, Silvia. *Caliban and the Witch.* [https://autonomedia.org/]: Autonomedia, 2004.

Ferdinand, Malcom. *Decolonial Ecology: Thinking from the Caribbean World.* Hoboken, NJ: John Wiley & Sons, 2021.

Ferguson, James. *The Anti-Politics Machine: "Development," Depoliticization and Bureaucratic Power in Lesotho.* [https://www.lib.cam.ac.uk/collections/departments/manuscripts-university-archives/subject-guides/university-archives-0-0]: CUP Archive, 1990.

Fletcher, Robert. *Failing Forward: The Rise and Fall of Neoliberal Conservation.* Oakland: University of California Press, 2023.

———. "How I Learned to Stop Worrying and Love the Market: Virtualism, Disavowal, and Public Secrecy in Neoliberal Environmental Conservation." *Environment and Planning D: Society and Space* 31, no. 5 (2013): 796–812.

———. "Neoliberal Environmentality: Towards a Poststructuralist Political Ecology of the Conservation Debate." *Conservation and Society* 8, no. 3 (2010): 171.

Fletcher, Robert, and Jan Breitling. "Market Mechanism or Subsidy in Disguise? Governing Payment for Environmental Services in Costa Rica," in "The Global Rise and Local Implications of Market-Oriented Conservation Governance," special issue, *Geoforum,* 43, no. 3 (May 2012): 402–11. https://doi.org/10.1016/j.geoforum.2011.11.008.

Fletcher, Robert, and Bram Büscher. "The PES Conceit: Revisiting the Relationship between Payments for Environmental Services and Neoliberal Conservation." *Ecological Economics* 132 (2017): 224–31.

Fletcher, Robert, Wolfram Dressler, Bram Büscher, and Zachary R. Anderson. "Debating REDD+ and Its Implications: Reply to Angelsen et al." *Conservation Biology* 31, no. 3 (2017): 721–23. https://doi.org/10.1111/cobi.12934.

———. "Questioning REDD+ and the Future of Market-Based Conservation." *Conservation Biology* 30, no. 3 (1 June 2016): 673–75. https://doi.org/10.1111/cobi.12680.

Forest Carbon Partnership Facility (FCPF). *Guyana Readiness Preparation Proposal (R-PP).* FCPF, 2012. https://www.forestcarbonpartnership.org/system/files/documents/FCPF%20-%20Readiness%20Preparation%20Proposal%20-%20Guyana%20December%202012.pdf.

Forsyth, Tim. *Multilevel, Multiactor Governance in REDD+: Participation, Integration and Coordination.* Bogor, Indonesia: Center for International Forestry Research (CIFOR), 2009. http://www.cifor.org/publications/pdf_files/books/bangelsen090209.pdf.

Foucault, Michel. *Power/Knowledge: Selected Interviews and Other Writings, 1972–1977.* New York: Pantheon, 1980.

———. *The Birth of Biopolitics: Lectures at the Collège de France, 1978–1979.* English translation. New York: Palgrave MacMillan, 2008.

Gebara, Maria Fernanda, and Arun Agrawal. "Beyond Rewards and Punishments in the Brazilian Amazon: Practical Implications of the REDD+ Discourse." *Forests* 8, no. 3 (2017): 66.

Glasgow, R. A. *Guyana: Race and Politics among Africans and East Indians.* Dordrecht: Springer Science & Business Media, 2012.

Goldstein, Daniel M. "Decolonialising 'Actually Existing Neoliberalism.'" *Social Anthropology* 20, no. 3 (2012): 304–9.

Government of Guyana. *A Low Carbon Development Strategy: Transforming Guyana's Economy While Combating Climate Change. Draft Report, Region 8, Kato.* (Consultation Report). 2009. http://www.lcds.gov.gy/index.php/documents/reports/local/consultations-reports/186-sub-national-consultation-draft-report-region-8-kato/file.

Government of Guyana. *Enhancing National Competitiveness: National Competitiveness Strategy (Draft).* 2006. http://finance.gov.gy/images/uploads/documents/ncs.pdf.

Graham, Don. "Flooding Slams Suriname; IMB Relief Effort Opens Doors for Sharing the Gospel." Baptist Press, 2006. http://www.bpnews.net/23271/flooding-slams-suriname-imb-relief—effort-opens-doors-for-sharing-the-gospel.

Green, Eric. "U.S. Providing Aid to Victims of Heavy Flooding in Suriname." IIP Staff Written. 12 May 2006. http://iipdigital.usembassy.gov/st/english/article/2006/05/20060512155437aeneerg0.1372492.html#axzz4FofFZUb3.

Greenpeace. *Bad Influence—How McKinsey-Inspired Plans Lead to Rainforest Destruction.* Greenpeace International. Accessed 14 December 2016. http://www.greenpeace.org/international/en/publications/reports/Bad-Influence/.

GRIF. *United Nations Development Programme Country: GUYANA Project Document.* 2014. http://www.guyanareddfund.org/images/stories/Signed%20ADF%20Phase%20II%20Project%20Document%20June%202014.pdf.

Griffiths, Tom, and Lawrence Anselmo. *Indigenous Peoples and Sustainable Livelihoods in Guyana: An Overview of Experiences and Potential Opportunities.* Forest Peoples Programme and North-South Institute, Moreton-in-Marsh, England, and Ottawa, Canada: 2010.

Griffiths, Tom, and Jean La Rose. *Searching for Justice and Land Security: Land Rights, Indigenous Peoples and Governance of Tenure in Guyana.* Forest Peoples Programme: 2014.

Grossberg, Lawrence. "On Postmodernism and Articulation: An Interview with Stuart Hall." *Journal of Communication Inquiry* 10, no. 2 (1986): 45–60.

Gupta, Tania Das, Carl E. James, Chris Andersen, Grace-Edward Galabuzi, and Roger C. A. Maaka. *Race and Racialization, 2E: Essential Readings.* Toronto: Canadian Scholars' Press, 2018.

Guyana Forestry Commission. *Guyana National Forest Plan.* 2011.

Guyana Lands and Surveys Commission. *Guyana National Land Use Plan.* 2013.

Guyana Ministry of Finance. *National Development Strategy.* 1997.

Haden, Philippa. *Forestry Issues in the Guiana Shield Region: A Perspective on Guyana and Suriname.* 1999. https://www.odi.org/sites/odi.org.uk/files/odi-assets/publications-opinion-files/5700.pdf.

Handler, Jerome S., and Matthew C. Reilly. "Contesting 'White Slavery' in the Caribbean: Enslaved Africans and European Indentured Servants in Seventeenth-Century Barbados." *New West Indian Guide / Nieuwe West-Indische Gids* 91, no. 1–2 (2017): 30–55.

Haraway, Donna. "Anthropocene, Capitalocene, Plantationocene, Chthulucene: Making Kin." *Environmental Humanities* 6, no. 1 (2015): 159–65.

Haraway, Donna J. *Staying with the Trouble: Making Kin in the Chthulucene.* Durham, NC: Duke University Press, 2016.

Harris, Paul G. *World Ethics and Climate Change: From International to Global Justice.* Edinburgh: Edinburgh University Press, 2009.

Harvey, David. *NeoLiberalism: A Brief History.* Oxford: Oxford University Press, 2005.

———. "Neo-Liberalism as Creative Destruction." *Geografiska Annaler: Series B, Human Geography* 88, no. 2 (2006): 145–58.

Heemskerk, Marieke. *Demarcation of Indigenous and Maroon Lands in Suriname.* Paramaribo, Suriname: Gordon and Betty Moore Foundation and Amazon Conservation Team Suriname, 2009.

———. *Rights to Land and Resources for Indigenous Peoples and Maroons in Suriname.* 2005. http://www.act-suriname.org/wp-content/uploads/2015/02/ACT_land-rights-report-2005.pdf.

Hoefte, Rosemarijn. *Suriname in the Long Twentieth Century: Domination, Contestation, Globalization.* Springer, 2013.

Hofman, Corinne, Angus Mol, Menno Hoogland, and Roberto Valcárcel Rojas. "Stage of Encounters: Migration, Mobility and Interaction in the Pre-Colonial and Early Colonial Caribbean." *World Archaeology* 46, no. 4 (2014): 590–609.

Hoogbergen, Wim, and Dirk Kruijt. "Gold, 'Garimpeiros' and Maroons: Brazilian Migrants and Ethnic Relationships in Post-War Suriname." *Caribbean Studies* 32, no. 2 (2004): 3–44.

Hook, Andrew. "Following REDD+: Elite Agendas, Political Temporalities, and the Politics of Environmental Policy Failure in Guyana." *Environment and Planning E: Nature and Space* 3, no. 4 (2019). https://doi.org/10.1177/2514848619875665.

———. "Mapping Contention: Mining Property Expansion, Amerindian Land Titling, and Livelihood Hybridity in Guyana's Small-Scale Gold Mining Landscape." *Geoforum* 106 (2019): 48–67.

Household Budget Survey. *Suriname Census 2012*. 2012.

Hout, Wil. "Development under Patrimonial Conditions: Suriname's State Oil Company as a Development Agent." *The Journal of Development Studies* 43, no. 8 (2007): 1331–50.

Hulme, David. *Global Poverty: Global Governance and Poor People in the Post-2015 Era*. London: Routledge, 2015.

Indufor, and Guyana Forestry Commission. *Guyana REDD+ Monitoring Reporting & Verification System (MRVS)*. Guyana Forestry Commission, 2014.

International Bank for Reconstruction and Development / The World Bank. *Suriname Sector Competitive Analysis: Identifying Opportunities and Constraints to Investment and Diversification in the Agribusiness and Extractives Sectors*. The World Bank Group, 2017. http://documents.worldbank.org/curated/en/928391488539939667/pdf/113150-ESW-v2-P156216-PUBLIC-Suriname-Sector-Competitiveness-Analysis-Report.pdf.

IPCC, 2007 [Bernstein, Lenny, Peter Bosch, Osvaldo Canziani, Zhenlin Chen, Renate Christ, Ogunlade Davidson, William Hare, et al.]. *Climate Change 2007: Synthesis Report. Contribution of Working Groups I, II and III to the Fourth Assessment Report of the Intergovernmental Panel on Climate Change*. Geneva: Intergovernmental Panel on Climate Change, 2007. https://www.ipcc.ch/report/ar4/syr/.

Jagan, Cheddi. *The West on Trial: My Fight for Guyana's Freedom* (rev. ed.). London: Hansib, (1966) 1997 .

Johnson, Melissa A. *Becoming Creole: Nature and Race in Belize*. New Brunswick, NJ: Rutgers University Press, 2018.

Jones, Cecily. "Contesting the Boundaries of Gender, Race and Sexuality in Barbadian Plantation Society." *Women's History Review* 12, no. 2 (2003): 195–232.

Josiah, Barbara. *Migration, Mining, and the African Diaspora: Guyana in the Nineteenth and Twentieth Centuries*. Springer, 2011.

Kambel, Ellen-Rose, and Fergus MacKay. *The Rights of Indigenous Peoples and Maroons in Suriname*. Copenhagen: International Work Group for Indigenous Affairs, 1999.

Kashwan, Prakash, Rosaleen V. Duffy, Francis Massé, Adeniyi P. Asiyanbi, and Esther Marijnen. "From Racialized Neocolonial Global Conservation to an Inclusive and Regenerative Conservation." *Environment: Science and Policy for Sustainable Development* 63, no. 4 (2021): 4–19.

Knight, Franklin W. *The Caribbean: The Genesis of a Fragmented Nationalism*. Oxford: Oxford University Press, 1990.

Larner, Wendy. "Neoliberalism?" *Environment and Planning D: Society and Space* 21 (2003): 509–12.

———. "Neo-Liberalism: Policy, Ideology, Governmentality." *Studies in Political Economy* 63, no. 1 (2000): 5–25.

Laws of Guyana, Forests Act, Chapter 67:01 § (2009).

Lemke, Thomas. "'The Birth of Bio-Politics': Michel Foucault's Lecture at the Collège de France on Neo-Liberal Governmentality." *Economy and Society* 30, no. 2 (2001): 190–207.

Li, Tania M. *Fixing Non-Market Subjects: Governing Land and Population in the Global South.* 2014. https://tspace.library.utoronto.ca/handle/1807/67595.

Li, Tania Murray. "Governmentality." *Anthropologica* 49, no. 2 (2007): 275–81.

Lowenthal, David. "The Range and Variation of Caribbean Societies." *Annals of the New York Academy of Sciences* 83, no. 1 (1960): 786–95.

Luke, Timothy W. "Tracing Race, Ethnicity, and Civilization in the Anthropocene." *Environment and Planning D: Society and Space* 28, no. 1 (2018). https://doi.org/10.1177/0263775818798030.

Maldonado-Torres, Nelson. "On the Coloniality of Being: Contributions to the Development of a Concept." *Cultural Studies* 21, no. 2–3 (2007): 240–70.

Malm, Andreas. *Fossil Capital: The Rise of Steam Power and the Roots of Global Warming.* London: Verso Books, 2016.

McGregor, Andrew, Edward Challies, Peter Howson, Rini Astuti, Rowan Dixon, Bethany Haalboom, Michael Gavin, Luca Tacconi, and Suraya Afiff. "Beyond Carbon, More than Forest? REDD+ Governmentality in Indonesia." *Environment and Planning A* 47, no. 1 (2015): 138–55.

McKinsey & Company. "Who We Are." Accessed 24 February 2017. http://www.mckinsey.com/about-us/who-we-are.

Menke, Jack, and Jerome Egger. *Country Report Suriname.* 2006.

Mhango, Baijah. "The Political Economy of Aid: The Case of Suriname." *Caribbean Studies* 24, no. 1/2 (1991): 123–64.

Mignolo, Walter D., and Catherine E. Walsh. *On Decoloniality: Concepts, Analytics, Praxis.* Durham, NC: Duke University Press, 2018.

Miller, Peter, and Nikolas Rose. "Governing Economic Life." *Economy and Society* 19, no. 1 (1990): 1–31.

Ministry of Labour, Technological Development and Environment. *Suriname Second National Communication to the United Nations Framework Convention on Climate Change.* 2013. http://www.nimos.org/smartcms/downloads/1.%20suriname%20snc-unfc%20on%20climatechange%20february%202013.pdf.

Mintz, Sidney Wilfred. *Sweetness and Power.* New York: Viking, 1985. http://www.followthethings.com/sweetnessandpower.shtml.

Mistry, Jay, Angela Berardi, Caspar Verwer, and Geraud Ville. *Up-Scaling Support for Community Owned Solutions: A Project Cobra Report for Policy Makers.* Cobra, 2015. https://www.researchgate.net/profile/Andrea-Berardi-4/publication/280839920_Up-scaling_support_for_community_owned_solutions/links/55c8dbb908aea2d9bdc91ede/Up-scaling-support-for-community-owned-solutions.pdf.

Mollett, Sharlene. "The Power to Plunder: Rethinking Land Grabbing in Latin America." *Antipode* 48, no. 2 (2016): 412–32.

Moore, Jason W. *Capitalism in the Web of Life: Ecology and the Accumulation of Capital.* London: Verso Books, 2015.

———. "Sugar and the Expansion of the Early Modern World-Economy: Commodity Frontiers, Ecological Transformation, and Industrialization." *Review (Fernand Braudel Center)* 23, no. 3 (2000): 409–33.

Mudge, Stephanie Lee. "What Is Neo-Liberalism?" *Socio-Economic Review* 6, no. 4 (2008): 703–31.

Myers, Rodd, Anne M. Larson, Ashwin Ravikumar, Laura F. Kowler, Anastasia Yang, and Tim Trench. "Messiness of Forest Governance: How Technical Approaches Suppress Politics in REDD+ and Conservation Projects." *Global Environmental Change* 50 (2018): 314–24.

Naughton-Treves, L., and C. Day. *Lessons about Land Tenure, Forest Governance and REDD+: Case Studies from Africa, Asia and Latin America.* Madison, Wisconsin: UW-Madison Land Tenure Center, 2012.

Neumann, Roderick P. *Imposing Wilderness: Struggles over Livelihood and Nature Preservation in Africa.* Oakland: University of California Press, 1998.

News Source Guyana. "Exxon Announces Its 15th Oil Discovery in Guyana." News Source Guyana, 23 December 2019. https://newssourcegy.com/news/exxon-announces-its-15th-oil-discovery-in-guyana/.

Norad. *Real-Time Evaluation of Norway's International Climate and Forest Initiative Synthesising Report 2007–2013.* Report 03/2014, 2014. https://www.norad.no/en/toolspublications/publications/2014/real-time-evaluation-of-norways-international-climate-and-forest-initiative.-synthesising-report-2007-2013/.

Office of the President, Guyana. "A Low-Carbon Development Strategy—Transforming Guyana's Economy While Combatting Climate Change." Policy Document. Office of the President, Guyana, 2010.

Office of the President, Republic of Guyana. "Creating Incentives to Avoid Deforestation." December 2008.

———. "Low Carbon Development Strategy Update." 2013. https://www.lcds.gov.gy/index.php/the-lcds/207-low-carbon-development-strategy-update-march-2013/file.

Offshore Energy Today. "Apache Cheers 'Significant Oil Discovery' Offshore Suriname." Offshore Energy, 7 January 2020. https://www.offshoreenergytoday.com/apache-cheers-significant-oil-discovery-offshore-suriname/.

Okereke, Chukwumerije, Harriet Bulkeley, and Heike Schroeder. "Conceptualizing Climate Governance beyond the International Regime." *Global Environmental Politics* 9, no. 1 (2009): 58–78.

Pachauri, Rajendra K., and Andy Reisinger. *IPCC Fourth Assessment Report.* Geneva: Intergovernmental Panel on Climate Change, 2007.

Pan American Health Organization. "Floods in Guyana—January/February 2005." 2005. http://www.paho.org/disasters/index.php?option=com_content&view=article&id=714%3Afloods-in-guyana-january%2Ffebruary-2005&catid=988%3Asmall-emergency-news&Itemid=0&lang=en.

Parker, Charlie, Andrew Mitchell, Mandar Trivedi, Niki Mardas, et al. *The Little REDD Book: A Guide to Governmental and Non-Governmental Proposals for Reducing Emissions from Deforestation and Degradation*. Oxford: Global Canopy Programme, 2008. https://www.cabdirect.org/cabdirect/abstract/20093053370.

Parreñas, Juno Salazar. *Decolonizing Extinction: The Work of Care in Orangutan Rehabilitation*. Durham, NC: Duke University Press, 2018.

Patterson, Orlando. *Slavery and Social Death: A Comparative Study, with a New Preface*. Cambridge, MA: Harvard University Press, 2018.

Peluso, Nancy Lee, and Peter Vandergeest. "Genealogies of the Political Forest and Customary Rights in Indonesia, Malaysia, and Thailand." *The Journal of Asian Studies* 60, no. 03 (2001): 761–812.

Peterson, Garry D., and Marieke Heemskerk. "Deforestation and Forest Regeneration Following Small-Scale Gold Mining in the Amazon: The Case of Suriname." *Environmental Conservation* 28, no. 02 (2001): 117–26.

Phelps, Jacob, Daniel A. Friess, and Edward L. Webb. "Win–Win REDD+ Approaches Belie Carbon–Biodiversity Trade-Offs." *Biological Conservation* 154 (2012): 53–60.

Prakash, Gyan. "Writing Post-Orientalist Histories of the Third World: Perspectives from Indian Historiography." *Comparative Studies in Society and History* 32, no. 02 (April 1990): 383–408. https://doi.org/10.1017/S0010417500016534.

Price, Richard. "The Maroon Population Explosion: Suriname and Guyane." *New West Indian Guide / Nieuwe West-Indische Gids* 87, no. 3–4 (1 January 2013): 323–27. https://doi.org/10.1163/22134360-12340110.

———. "Uneasy Neighbors: Maroons and Indians in Suriname." *Tipití: Journal of the Society for the Anthropology of Lowland South America* 8, no. 2 (1 December 2010). http://digitalcommons.trinity.edu/tipiti/vol8/iss2/4.

Quijano, Anibal. "Coloniality of Power and Eurocentrism in Latin America." *International Sociology* 15, no. 2 (2000): 215–32.

Rabe, Stephen G. *US Intervention in British Guiana: A Cold War Story*. Chapel Hill: University of North Carolina Press, 2006.

Rattansi, Ali. "Postcolonialism and Its Discontents." *Economy and Society* 26, no. 4 (1 November 1997): 480–500. https://doi.org/10.1080/03085149700000025.

Read, Jason. "A Genealogy of Homo-Economicus: Neoliberalism and the Production of Subjectivity." *Foucault Studies* 6 (February 2009): 25–36.

Reporter, Staff. "Amerindian Tribes of Guyana." *Guyana Chronicle* (blog), 17 September 2010. https://guyanachronicle.com/2010/09/17/amerindian-tribes-of-guyana-2.

Requena-i-Mora, Marina, and Dan Brockington. "Seeing Environmental Injustices: The Mechanics, Devices and Assumptions of Environmental Sustainability Indices and Indicators." *Journal of Political Ecology* 28, no. 1 (2021).

Richardson, David, and David Eltis. *Atlas of the Transatlantic Slave Trade*. New Haven, CT: Yale University Press, 2015.

Robinson, Cedric J. *Black Marxism, Revised and Updated Third Edition: The Making of the Black Radical Tradition*. Chapel Hill: University of North Carolina Press, 2020.

Rodney, Walter. *A History of the Guyanese Working People, 1881–1905*. Baltimore: Johns Hopkins Press, 1981.

———. *How Europe Underdeveloped Africa*. London: Verso Trade, 2018.

Saldanha, Arun. "Race." In *The Wiley-Blackwell Companion to Human Geography*, edited by J. A. Agnew and J. S. Duncan, 453–64. Hoboken, NJ: John Wiley & Sons, 2011. https://doi.org/10.1002/9781444395839.ch32.

Scott, James C. *Seeing Like a State: How Certain Schemes to Improve the Human Condition Have Failed*. New Haven, CT: Yale University Press, 1998.

Shohat, Ella. "Notes on the 'Post-Colonial.'" *Social Text*, no. 31/32 (1992): 99–113. https://doi.org/10.2307/466220.

Simon, David. "Separated by Common Ground? Bringing (Post) Development and (Post) Colonialism Together." *The Geographical Journal* 172, no. 1 (2006): 10–21.

Singh, Chaitram. "Re-Democratization in Guyana and Suriname: Critical Comparisons." *European Review of Latin American and Caribbean Studies* 84 (2008): 71.

Sizer, Nigel, and Richard Price. *Backs to the Wall in Suriname: Forest Policy in a Country in Crisis*. Washington, DC: World Resources Institute, 1995. https://www.wri.org/publication/backs-wall-suriname.

Skutsch, Margaret, and Esther Turnhout. "How REDD+ Is Performing Communities." *Forests* 9, no. 10 (2018): 638.

Stabroek News. "Shuman Initially Prevented from Attending High Court in Indigenous Attire." Stabroek News, 11 March 2020. https://www.stabroeknews.com/2020/03/11/news/guyana/shuman-initially-prevented-from-attending-high-court-in-indigenous-attire/.

Stearman, Allyn MacLean. "Revisiting the Myth of the Ecologically Noble Savage in Amazonia: Implications for Indigenous Land Rights." *Culture & Agriculture* 14, no. 49 (1 March 1994): 2–6. https://doi.org/10.1525/cuag.1994.14.49.2.

Stern, Nicholas. *The Economics of Climate Change: The Stern Review*. Cambridge: Cambridge University Press, 2007.

Struiken, Harold, and Chris Healy. "Suriname: The Challenge of Formulating Land Policy." In *Land in the Caribbean: Issues of Policy, Administration and Management in the English-Speaking Caribbean*, edited by Allen N. Williams, 315–44. Mount Horeb, WI: Terra Institute, 2003.

Sundberg, Juanita. "Placing Race in Environmental Justice Research in Latin America." *Society and Natural Resources* 21, no. 7 (2008): 569–82.

Suriname Readiness Preparation Proposal (R-PP). FCPF, 2013. https://www.forestcarbonpartnership.org/system/files/documents/Suriname_R-PP-finaldraft_23Feb.pdf.

Svensson, Sara. *(Suriname) National Plan for Forest Cover Monitoring*. ONF International, 2014. https://sbbsur.com/wp-content/uploads/2015/06/Forest_Cover_Monitoring_Plan_FCMP_Suriname.pdf.

The Private Sector Commission of Guyana, Conservation International-Guyana, and World Wildlife Fund (Guianas). "Bridging Guyana's Mining Sector with Its Low Carbon Development Path: Stimulating Multi-Stakeholder Dialogue—Facilitating Convergence." Stakeholder forum report. 2013.

Thompson, Alvin O. *Flight to Freedom: African Runaways and Maroons in the Americas*. University of the West Indies Press, 2006.

Trafford, James. "Empire's New Clothes: After the 'Peaceful Violence' of Neoliberal Coloniality." *Angelaki* 24, no. 1 (2019): 37–54.

Tuck, Eve, and K. Wayne Yang. "Decolonization Is Not a Metaphor." *Decolonization: Indigeneity, Education & Society* 1, no. 1 (2012).

Tuhiwai Smith, Linda. "Decolonising Methodologies, 20 Years On." Presented at the The Sociological Review Annual Lecture, London, UK, 16 October 2019. https://www.youtube.com/watch?v=YSX_4FnqXwQ.

United Nations, Climate Change. "What Is REDD+? | UNFCCC." Accessed 15 November 2022. https://unfccc.int/topics/land-use/workstreams/redd/what-is-redd.

Van Hecken, Gert, Vijay Kolinjivadi, Catherine Windey, Pamela McElwee, Elizabeth Shapiro-Garza, Frédéric Huybrechs, and Johan Bastiaensen. "Silencing Agency in Payments for Ecosystem Services (PES) by Essentializing a Neoliberal 'Monster' into Being: A Response to Fletcher & Büscher's 'PES Conceit.'" *Ecological Economics* 144 (2018): 314–18.

van Kuijk, Marijke. *REDD+ Development in the Guianas: The Evolution of the Concept and Activities Undertaken in Guyana, Suriname and French Guiana*. WWF Guianas, 2012. http://d2ouvy59p0dg6k.cloudfront.net/downloads/redd__developments_in_the_guianas_2.pdf.

Vandergeest, Peter, and Nancy Lee Peluso. "Political Forests." In *The International Handbook of Political Ecology*, edited by Raymond L. Bryant, chapter 12. Cheltenham, UK: Edward Elgar Publishing, 2015.

Vaughn, Sarah E. "Reconstructing the Citizen: Disaster, Citizenship, and Expertise in Racial Guyana." *Critique of Anthropology* 32, no. 4 (2012): 359–87.

Vezzoli, Simona. *The Evolution of Surinamese Emigration across and beyond Independence: The Role of Origin and Destination States.* Working Papers. Oxford: International Migration Institute, University of Oxford, 2014. https://www.migrationinstitute.org/publications/wp-106-2014/@@download/file.

Wertz-Kanounnikoff, Sheila, Louis V. Verchot, M. Kanninen, and D. Murdiyarso. "How Can We Monitor, Report and Verify Carbon Emissions from Forests?" In *Moving Ahead with REDD: Issues, Options and Implications,* edited by Arild Angelsen, chapter 9. Bogor, Indonesia: Center for International Forestry Research (CIFOR), 2008. http://www.cifor.org/publications/pdf_files/Books/BAngelsen080109.pdf.

West, Paige, and Dan Brockington. "Introduction: Capitalism and the Environment." *Environment and Society* 3, no. 1 (1 September 2012): 1–3. https://doi.org/10.3167/ares.2012.030101.

Whyte, Kyle. "Settler Colonialism, Ecology, and Environmental Injustice." *Environment and Society* 9, no. 1 (1 September 2018): 125–44. https://doi.org/10.3167/ares.2018.090109.

Williams, Eric. *Capitalism and Slavery.* Chapel Hill: University of North Carolina Press, (1944) 1994.

Wolfe, Patrick. *Traces of History: Elementary Structures of Race.* London: Verso Books, 2016.

World Bank. "Suriname Overview." September 2016. http://www.worldbank.org/en/country/suriname/overview.

WWF. *Community-Based Monitoring, Reporting and Verification Know-How: Sharing Knowledge from Practice.* WWF, 2015. https://wwfint.awsassets.panda.org/downloads/cmrv_web.pdf.

WWF. "FCPF Approves Suriname's REDD+ Readiness Preparation Proposal (R-PP) with Grant for US$3.8m." WWF, 2013. http://wwf.panda.org/wwf_news/?209337/FCPF-Approves-Surinames-REDD-Readiness-Preparation-Proposal-R-PP-with-grant-for-US38m.

Wynter, Sylvia. "Unsettling the Coloniality of Being/Power/Truth/Freedom: Towards the Human, after Man, Its Overrepresentation—An Argument." *CR: The New Centennial Review* 3, no. 3 (2003): 257–337.

Yusoff, Kathryn. *A Billion Black Anthropocenes or None.* Minneapolis: University of Minnesota Press, 2018.

Index

Note: References in *italic* and **bold** refer to figures and tables. References followed by "n" refer to endnotes.

Founded in 1893,
UNIVERSITY OF CALIFORNIA PRESS
publishes bold, progressive books and journals on topics in the arts, humanities, social sciences, and natural sciences—with a focus on social justice issues—that inspire thought and action among readers worldwide.

The UC PRESS FOUNDATION
raises funds to uphold the press's vital role as an independent, nonprofit publisher, and receives philanthropic support from a wide range of individuals and institutions—and from committed readers like you. To learn more, visit ucpress.edu/supportus.